Pannell Family

DEDICATION

To my long-suffering family (particularly my wife Sandra) who have stayed with and supported me through this tortuous journey.

ONCE MORE UNTO THE VINE

BY

DR. WILLIAM PANNELL A.M.

Once More Unto the Vine

Published in Australia by Rocket Frog Books 2011
A Division of Rigby Transmedia
6256 Caves Road, Margaret River, Western Australia 6285
Produced and published at the request and permission of the Dr. William Pannell A.M.

Design, image editing and production: Rigby Transmedia
Editing and content management: Peter Rigby and Jodie Pannell
Primary photography: Jodie Pannell
Printing: Lamb Print

ISBN: 978-0-9804437-6-9

FOREWARD
BY HUON HOOKE

In the 'old days' in France, it was accepted wisdom that it took three generations to build up a great wine estate. These days, people like Bill Pannell manage to do it more than once in their own lifetime. It takes an exceptional person to achieve this, of course, and Bill is such a man.

Moss Wood is well established as one of Australia's, nay the world's, great wineries. Picardy is still emerging but has already produced some truly excellent wine and I have no doubt it has the potential to be great.

Bill is not only a man of energy, ambition and intelligence, he is also a quick learner, and has enough humility to realise when someone else knows better. Hence, he has been a frequent visitor to France and an avid student of French ways.

If there's one thing that makes him different from many other Australians who've achieved great things in wine, it's that he understood that several centuries and many generations worth of experience is something that shouldn't be dismissed as outdated, hidebound or arrogant – which many Australians have done of the French, especially in matters viticultural.

I've known and admired Bill for nearly 30 years, and what this book revealed to me for the first time is that he's also a very good writer. I commend his book to you.

I have for some time been under pressure from my family and friends to record those events relating to my lifelong affair with wine.

Browsing through various bookshops, trying to develop a feel for style and format, one cannot help but be impressed by the number of biographies currently being produced.

Two prominent Australian politicians with aspirations to leadership, but certainly several years away from the realization of their ultimate goals, have already had their biographies published.

Unlike them, I can claim to be nearing the end of my journey. I will most certainly never plant another vineyard or build another winery. That being the case, I feel justified in putting pen to paper in order to document my life's work.

Frankly, I find the litany of those worthy of recognition which generally precedes such publications rather banal.

So to my wife, who has been at my side throughout this saga and to my children, who undoubtedly endured considerable hardship during those difficult early years, I offer special tribute.

To the legion of friends and supporters who have helped us to realise our dreams: whoever you are, wherever you are, thank you.

I ascribe very strongly to the notion that brevity is the soul of wit. "The Old Man and The Sea" is the only piece of fiction which I have enjoyed reading from cover to cover. Accordingly, I have tried to make this work as relevant and concise as possible.

- William Pannell

Bill, Sandra, Jodie & Dan

CHAPTER 1

INSPIRATION

Bill MacDonald was no ordinary teacher. Like the Jesuits who had sculptured both my soul and intellect in its formative years, the UWA professor of paediatrics believed that "book learning" was but one component of a well rounded education. The professor had a well developed interest in the vinous product of the grape and on more than one occasion, during our period of attachment to his unit (in an undergraduate capacity), we were rounded up and transported to the cellars of Houghton Wines in the nearby Swan Valley.

Legendary winemaker Jack Mann

The winemaker at Houghton at that time was, and had been since 1922, the legendary Jack Mann. Jack's passions in life were heralded by his dictum of "the three Cs": Christianity, Chablis and cricket. Jack was, during his many years at Houghton, a stalwart promoter of wine and its virtues. It would, I believe, be true to say that the general level of interest and erudition in fine wines in Western Australia owes much to Jack's seminal influence.

My first encounter with the great man occurred in 1964. Some ten years previously, an exchange scholar from Davis University in California, Dr John Olmo, had spent a year in Perth, ostensibly to examine the potential for production of fine table wine in the Swan Valley. He fairly rapidly concluded that the Swan Valley was rather too warm to produce the more delicate styles of table wine. He reasoned, however, that in a state the size of Western Australia, more suitable locations must exist.

Olmo applied temperature summation parameters which had been developed by one of his Californian mentors (Winkler) and determined that an area to the north of Albany and to the west of Mount Barker should be eminently suited to fine table wine production.

On the basis of Olmo's predictions, an investigatory committee was formed to examine suitable sites for vine trial plots in the Mount Barker region. Jack Mann was a member of that committee.

During our several visits to Houghton's cellars, Jack discussed

The present day Western Australian wine regions. In the early 1960s the focus was still very much on the Swan Valley near Perth and the current Geographic Indicators for the southern regions didn't exist.

with us his conviction that southern areas of Western Australia would soon produce table wines of a quality comparable to the worlds finest. As an avid West Australian, Jack wasn't given to the use of diminutives in such matters. This struck me, at the time, as a most exciting prospect. Indeed, it was to play a major role in shaping my destiny.

Several years later, as a young married medical graduate, on the occasion of our first wedding anniversary, I took my wife Sandra to one of Perth's better restaurants for a celebratory dinner. On the recommendation of the sommelier, we ordered a bottle of Penfold's newly released Grange Hermitage (vintage 1961 as I recall). My recollection is that the first taste of that wine was a defining moment in my life. Such richness! Such complexity! Such balance! I believe it was at that point that I determined that somehow, I would make a wine of comparable quality.

The following year (1967) we moved to Melbourne and took the opportunity to research the Eastern States wine scene in some detail. Among others, we visited the Purbricks at Chateau Tahbilk, John Brown at Milawa and Max Lake in the Hunter Valley, all of whom were encouraging (in varying degrees) with regard to my ambition to found a boutique vineyard and winery in South Western Australia. Importantly, we called on the vineyard manager of Seppelts new vineyard at Drumborg, near Portland in Victoria, Ross Heinze. Ross had actually toured Western Australia, looking for potential vineyard sites and confided his preference for the Cowaramup area, to the north of Margaret River, for red wine production. He believed that Mount Barker was better suited to white wine production.

Being as I was more interested in red wine and as my parents had a holiday home in Dunsborough (not far from Cowaramup), this particular piece of counsel suited me rather nicely!

I had, in the meantime, contacted the Western Australian Department of Agriculture, seeking information with respect to the Margaret River/Cowaramup area. I was forwarded a paper by Dr John Gladstones, (UWA graduate, plant breeder and wine enthusiast). Gladstones had spent his early years in the area around Margaret River and had an intuition that it would be suited to table wine production. He applied Winkler's parameters to the region (together with some of his own) and concluded that the area was, indeed, most suitable.

The Grange Hermitage 1961 - The great vintage that inspired me to try making my own wines.

CHAPTER 2

SEEK AND YE SHALL FIND

Tom Cullity - pioneer of the Margaret River wine industry

On our return to Western Australia in 1968, using the Dunsborough cottage as a base, we conducted a very thorough survey of the area between Busselton and Margaret River.

What little I knew at the time suggested that we should be looking for well-drained gravelly slopes. It fairly rapidly became clear that these were largely confined to the granite-based ridge running between Cape Naturaliste in the north and Cape Leeuwin in the south. Gladstones had suggested that temperatures to the south of Margaret River could be undesirably low, so we confined our search to those portions of the Leeuwin-Naturaliste ridge which were to the north of Margaret River.

For many weeks, we traversed every road and track in our target area: leaping out of the car with a shovel to dig a hole in anything which looked at all promising. Imagine our surprise when, on a small side road to the west of the town of Cowaramup, we came across a small, recently planted vineyard. We learned that a Perth cardiologist, Dr Tom Cullity had been similarly inspired by predictions of the region's potential and had already planted five acres of vines. He had called his vineyard Vasse Felix.

Willyabrup today - The cradle of early wine production in the Margaret River region. It looks a lot different today than when we started out in the late 1960s.

Eventually, in a rather pretty and secluded valley, some four kilometres to the north of Cullity's vineyard and in the drainage basin of the Willyabrup Brook, we found what we were looking for. On either side of a stream running from west to east across the property were gently sloping gravelly loams overlying retentive yellow clays.

A few discreet inquiries revealed that the property belonged to a local farmer by the name of Jack Guthrie. Interestingly though, it had originally been settled by an American seaman, Kinsella, who had jumped ship and (presumably) gone to this rather remote corner of the country in order to avoid detection. More recently, it had been owned by a rather reclusive German immigrant named Rapse.

A direct approach to the Guthries resulted, after some discussion "en famille", in agreement for me to subdivide off some twenty six acres of the most suitable soils. At that time, I was still working as a hospital resident (on wages which today would be considered immoral) and raising a family of two young sons, Stephen and Daniel. Funds were, therefore, rather scarce and we could not afford to purchase a large property.

Given that we were committed to red wine production and that Shiraz was then considered a rather "common" variety, we decided to plant Cabernet Sauvignon. Jack Mann used to refer to this variety as "the only grape allowed in heaven". True to form, Jack was very supportive and organized his vineyard workers at Houghton to take the required cuttings for us.

Houghton was one of the first and most significant vineyards planted in Western Australia. As such, it remains the ultimate source of much of the vine material planted in this state. The material used to plant Houghton itself was probably obtained from Europe just prior to the outbreak of phylloxera on the continent. It possibly represents a unique reliquary of material which was subsequently lost in Europe.

Our precious cuttings were planted in a nursery on our new

property in the spring of 1969. A year later, having cleared a five acre patch of both stumps and stones and surrounded it with a rabbit-proof fence, we commenced planting our first vineyard. Using a homemade planting wire, with beads of solder every six feet (1.8 metres), I dug a substantial hole for each vine and Sandra (by then heavily pregnant with our third son Nicholas) compacted each plant in to its chosen position using a heavy lead ramming tool.

In due course, jarrah posts were positioned in the vine rows with heavy jarrah strainers at the ends of each row. The posts were drilled and wires were drawn through each post and secured to the strainers.

All of this development involved quite a deal of seriously hard work, but it was not without its lighter moments. One of my more enduring memories is of Sandra, near-term with Nicholas and drenched with insecticide from a ruptured spray line, vaulting a barbed wire fence and shedding clothing in her desperate urgency to immerse herself in the nearby creek.

Our two sons, Stephen and Daniel, would spend many hours in the creek, constructing dams and catching yabbies. Like most boys of their age, they seemed impervious to cold and would frolic, near naked in the freezing water of the stream, even on the bleakest of winter days.

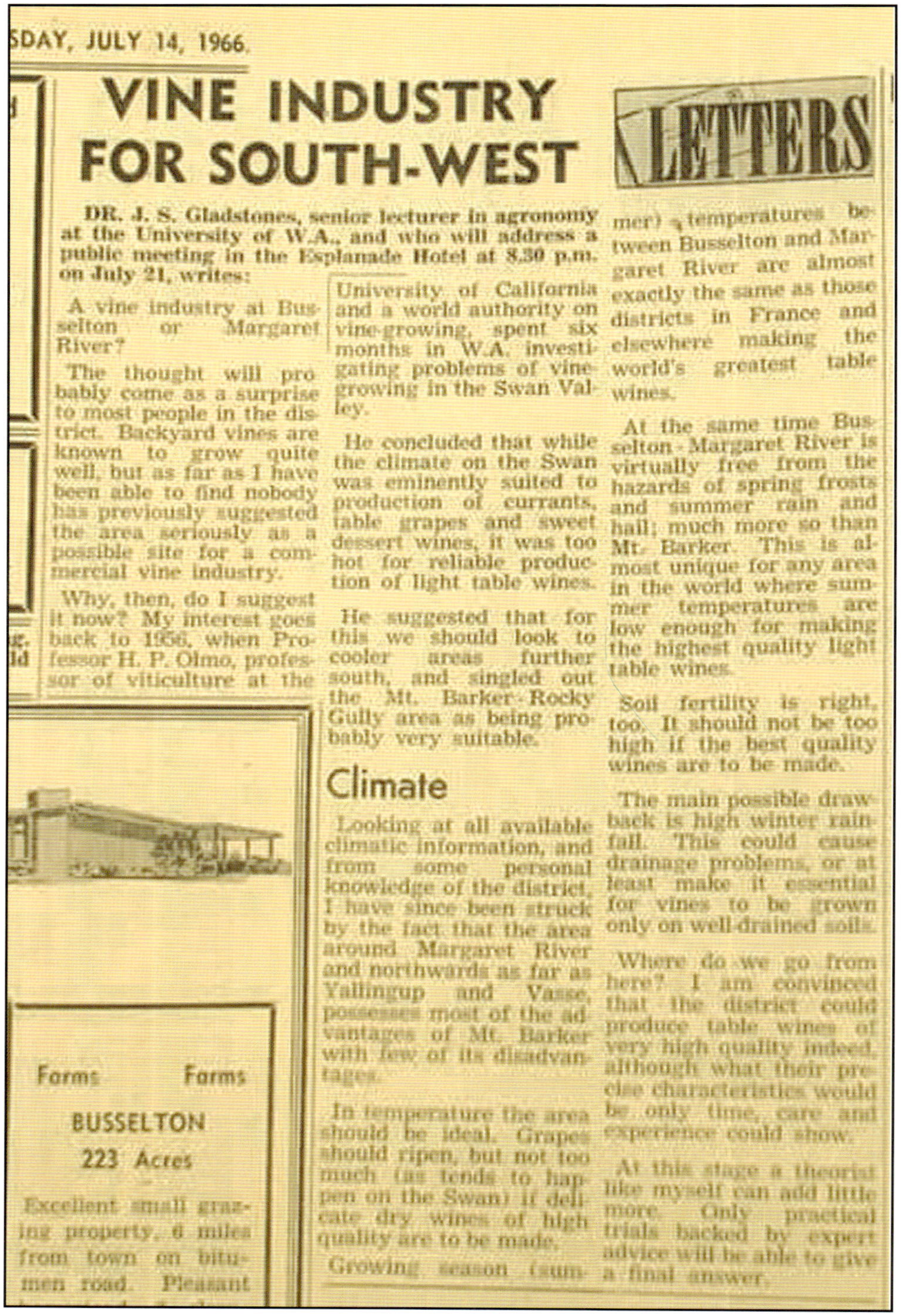

SDAY, JULY 14, 1966.

VINE INDUSTRY FOR SOUTH-WEST

DR. J. S. Gladstones, senior lecturer in agronomy at the University of W.A., and who will address a public meeting in the Esplanade Hotel at 8.30 p.m. on July 21, writes:

A vine industry at Busselton or Margaret River?

The thought will probably come as a surprise to most people in the district. Backyard vines are known to grow quite well, but as far as I have been able to find nobody has previously suggested the area seriously as a possible site for a commercial vine industry.

Why, then, do I suggest it now? My interest goes back to 1956, when Professor H. P. Olmo, professor of viticulture at the University of California and a world authority on vine-growing, spent six months in W.A. investigating problems of vine-growing in the Swan Valley.

He concluded that while the climate on the Swan was eminently suited to production of currants, table grapes and sweet dessert wines, it was too hot for reliable production of light table wines.

He suggested that for this we should look to cooler areas further south, and singled out the Mt. Barker-Rocky Gully area as being probably very suitable.

Climate

Looking at all available climatic information, and from some personal knowledge of the district, I have since been struck by the fact that the area around Margaret River and northwards as far as Yallingup and Vasse, possesses most of the advantages of Mt. Barker with few of its disadvantages.

In temperature the area should be ideal. Grapes should ripen, but not too much (as tends to happen on the Swan) if delicate dry wines of high quality are to be made.

Growing season (summer) temperatures between Busselton and Margaret River are almost exactly the same as those districts in France and elsewhere making the world's greatest table wines.

At the same time Busselton-Margaret River is virtually free from the hazards of spring frosts and summer rain and hail; much more so than Mt. Barker. This is almost unique for any area in the world where summer temperatures are low enough for making the highest quality light table wines.

Soil fertility is right, too. It should not be too high if the best quality wines are to be made.

The main possible drawback is high winter rainfall. This could cause drainage problems, or at least make it essential for vines to be grown only on well-drained soils.

Where do we go from here? I am convinced that the district could produce table wines of very high quality indeed, although what their precise characteristics would be only time, care and experience could show.

At this stage a theorist like myself can add little more. Only practical trials backed by expert advice will be able to give a final answer.

LETTERS

Farms Farms
BUSSELTON
223 Acres
Excellent small grazing property, 6 miles from town on bitumen road. Pleasant

An article in the Busselton-Margaret River Times dated July, 1966 espousing the region's potential for premium wine production. Dr Gladstones was right on target!

Sandra with our young family

Financially speaking, things were getting a little easier and we were able to start building our first house in Busselton. Nevertheless, we had to watch the pennies and would never miss an opportunity to supplement the larder with rabbit or wild duck shot on the property. During the 1971 - 1972 season, we trained our first vines to the wire and in the ensuing season it became obvious that we would soon produce a crop of grapes. We had now to consider what to do with them.

CHAPTER 3

STONE UPON STONE

The visitor to the Margaret River region today will find it difficult to imagine the degree of isolation which prevailed in the wine industry in the area in those early years. Nowadays it is simple enough to find someone to process the grapes which one has produced. In the seventies, there was no such option. We simply had to build a winery!

Having recently built a house for our young family and purchased our share of a medical practice, we were hardly awash with funds.

The perception in the region was, however that all doctors were, as if by definition, rich. Accordingly, our price for posts, labour, etc, was at a substantial premium to that applying to other land holders in the area.

To keep costs under control, therefore, with the help of a local artisan, I constructed a small brick building with a fibro-cement roof; barely adequate in size to process our first small vintage of five tons.

The rule of thumb applying to winery construction is to think of the largest building you might possibly need and double its size. In this case it was a matter of designing the smallest building we could possibly get away with and building exactly that!

It was as I recall at about that time that my parents paid a visit to the vineyard. My mother commented that the approach to the winery through a grove of ancient peppermint trees reminded her very much of the estate of a friend of hers in England. The name of that estate was Moss Wood. Sandra and I had spent many fruitless hours trying to think of a name for the property. When my mother spoke that name, we instantly realised that our search was over.

Our first small vintage, in 1973, was crushed and destemmed by our one extravagance - a Celestin Coq "fouloir/egrappoir". This machine was, by today's standards, of rather primitive construction but had the distinct advantage (for its time) of being able to crush the grape berries without damaging the pips and stalks.

The crushed grapes were fermented in concrete cattle troughs, lined with paraffin wax and pressed with a tiny hand-basket press obtained from an Italian food importer in Perth. The pressed juice was run directly into American oak barrels; French oak being virtually unobtainable in those days.

The new wine was obviously young and raw, but seemed to us (and to others who tasted it) to be surprisingly fruity and drinkable. In view of this, we were somewhat taken aback when a wine consultant and guru from Sydney tasted the wine and stated that it "tasted like duck's guts"!

It took us a few years to learn that it was almost de rigeur for the wise men of wine from the mystic east to cast disparaging remarks upon anything from Western Australia. Eventually, we developed a practice of giving them a glass of "Hunter Valley tar" or "McLaren Vale cow pats", on the blind, before showing them our wine. This seemed to have the effect of making them a little more guarded in their comments by the time they tasted our produce!

In 1974, a cask room was added to the initial fermentation cellar. In that same year, a further eight acres of Cabernet Sauvignon and 240 vines of Semillon were planted. I felt intuitively that the Margaret River area was unsuited to Rhine Riesling which others in the area were planting at that time but the WA Department of Agriculture did have substantial supplies of Semillon cuttings. I felt it was worth a try!

The 1974 vintage was slightly larger than its predecessor and produced tight wines of great colour and nervosity. A fairly basic cooling system, employing a refrigerated stainless steel coil which could be lowered into the fermenters, had been installed ready for the 1974 vintage.

The growing season of 1974-75 was atypically cool and I recall picking the Cabernet, at 12.5 degrees Baume ("under-ripe" by today's standards) in drizzling rain. The wine was fermented in newly installed stainless steel vessels and stored, prior to going into barrel in a stainless steel tank. We were able to obtain some French Oak barrels for this vintage as a South African order had fallen through and the surplus barrels had been sent on to Australia.

Planting the first vines at Moss Wood with Sandra, Stephen and Daniel in 1970

The 1975 Cabernet was, I believe, the best wine which I produced at Moss Wood. From the outset it showed tremendous fruit, great finesse and impeccable balance. As with all great vintages, one had the sensation of being simply an observer; standing by while some supernatural influence transformed the grapes into superlative wine.

A visiting Master of Wine, Edmund Penning - Rowsell, on tasting the 1975 Cabernet, pronounced it one of the best young wines of its type which he had tasted outside Bordeaux. After the "duck's guts" comment of the Sydney guru, this was exactly the sort of praise and support we required to restore our battered confidence.

Our attitude to marketing in those early days was pretty much as it is now. We felt that if we could make the wine as well as we possibly could and then sell it at a reasonable price, then the customers would come to us. In retrospect, there wasn't a lot of high quality wine about in those days and there was great interest in the newly emerging wine area of Margaret River. For whatever reason, we didn't seem to have any difficulty in finding a home for our produce.

As I have already said, we considered the 1975 Cabernet to be one of the best wines we have ever made. The occasional bottle which we open still drinks superbly. Although it had beautiful flavour, it was still tight and rather acid as a youngster. About a fortnight after we had released it, we received a call from a customer to say that he had purchased a case of the wine and found it rather acid. Nevertheless, he had managed to drink the whole case and did we have any more?

It was at about that time that our neighbour, Ellis Butcher, who had been working on our boundary fence, adjacent to the vineyard, asked me if I could show him how to prune vines. I invited him to hop the fence and join me and we spent the rest of the afternoon pruning together. Thus began a long association, with Ellis's wife Margaret coming to work for us on Moss Wood and quite rapidly assuming the role of "foreperson" in the vineyard and winery. Ellis developed a burning interest in viticulture and has been responsible for much of the vineyard development which has occurred in the Willyabrup region since that time. The Butchers are warm, genuine and industrious people who remain firm friends to this day. (At the time of publication we regret to say that Ellis has left us for “the great vineyard in the sky”.)

In searching for someone to help with pruning, I came across an elderly Italian, who lived not far from Moss Wood, by the name of Giovanni Palandri. "Old John", (as he came to be known) was the type of European who, despite his lack of formal education, was both worldly wise and cultivated. John had worked in vineyards in Italy, Corsica and France. His language was a blend of his native, northern Italian dialect and "timber-mill English" (English with fifty percent expletives) which became known, at Moss Wood as "Palandri". John had worked in timber mills for most of his time in Australia. He had recently lost his wife and was, I believe, glad to become involved in vineyard work and to have the company of others which that entailed. John's constant flow of anecdotes from "the old country", delivered in his rather unique patois, often had us paralysed with mirth as we worked our way along the vine rows.

One day, during the pruning season, in 1975, we were paid a visit by a Royal Australian Navy Helicopter, which landed in an open paddock next to the winery. Its mother-ship had been cruising off the coast nearby and one of the crew had heard that Moss Wood was making good wine, so they sent the helicopter crew to fetch some!

The 1975 - 76 season was extremely warm and much of the Cabernet was very ripe when it was picked in near heat-wave conditions. The combination of ripe, sweet fruit, fairly high alcohol levels (by the standards of the time, but certainly not by today's standards) together with a mix of French and American

oak, produced a wine which drove the show judges wild. It won a gold medal in open classes at the Perth Wine Show and was the first wine from Margaret River to achieve this distinction. A glass of the wine (held by Sandra) was featured on the front page of *The West Australian*. From that point onward, Moss Wood became a household word in Western Australia.

Be that as it may, I have always preferred the style of the 1975 Cabernet to that of the 1976.

The 1976 vintage was also the first to involve Gill Thomas. Gill's wife Jan, who was a general practitioner, agreed to do a locum for me while I attended to vintage. Gill, a Bunbury pharmacist, was also a keen wine enthusiast and wanted to be involved in a vintage. He brought his caravan to Moss Wood and parked it next to the winery so that he could be on the spot. Gill's presence was an enormous comfort to me during the long nights of plunging ferments and cooling tanks. During the day, he picked grapes and gathered the full picking buckets. At the winery, he helped feed the grapes through the crusher and with the necessary but arduous "clean up" at the end of each day.

Our picking team consisted mainly of farmers' wives. When the team began to flag at the end of a long picking session, Gill's comments with regard to "whingeing bloody farmers" was guaranteed to produce a burst of energy which would see buckets filled in record time.

Gill was a regular participant in the next eight vintages at Moss Wood. I remember that shared experience as a time of considerable happiness and satisfaction. In retrospect, I doubt if the vintages in question would have been so successful without Gill's support.

The 1977 vintage was unremarkable, producing sound, ripe wines from a fairly warm year. We also produced a small quantity of Pinot Noir in that year. The only sour note was that some of the barrels supplied to us seemed a little "green" or under-seasoned. Despite this, the wood flavour seems to have "ripened - up" in bottle and the wine is generally accepted to be up to Moss Wood's usual level of quality.

Despite growing quite a large crop of excellent quality fruit, the 1978 vintage was a near disaster for us.

As a rule, we obtained our yeast cultures from the WA Department of Agriculture. That which we preferred was a yeast known as WRI729. The yeast which was supplied to us in 1978 was labelled WRI729A. I was assured that it was simply a sub-culture of the original WRI729 and used it without making enquiries as to its origins. In fact, WRI729A was a sub-culture which had been re-imported from South Australia. At some stage, this particular yeast had mutated and produced prodigious amounts of hydrogen sulphide (rotten egg gas). Our precious Cabernet smelled more like an old swamp than a fine claret!

We eventually employed a technique known as copper fining to remove the sulphide from the wine. I was far from happy with the end result and rather than jeopardize the good name that we had worked so hard to establish, decided to sell it as a "dry red" at a ridiculously low price. In fact, the sacrifice served only to strengthen our reputation for both quality and integrity. At that time though, from the financial point of view, it was exactly what we didn't need!

By the end of 1978, we had nearly twenty acres under vine. I was running a busy medical practice during the day and delivering babies at night. Vineyard and winery activities such as pruning, winemaking and bottling were "squeezed" in on weekends, evenings and holidays. I had not actually had a real holiday for over five years. As a consequence, I had become physically and mentally exhausted. I soon realised that I could simply not carry on as I had been doing. In short, I needed help!

The tribe!

CHAPTER 4

KEITH

Bill Hardy was a member of the famous South Australian wine making family. He had been installed as winemaker at Houghton shortly after Hardys purchased that enterprise in 1976.

Bill was, not unnaturally, interested in developments in the Margaret River region and he began to visit us as Moss Wood on a fairly regular basis. I had great respect for Bill as a winemaker. I found him to be genial and affable and he was highly supportive of our early efforts at Moss Wood.

French oak barrels at Moss Wood

Having determined to seek qualified assistance in the vineyard and winery, I asked Bill if he knew of anybody who would be interested in such a position. It so happened that the son of a South Australian medico, who was a close friend of the Hardy family, was on the point of completing his degree in viticulture and oenology at Roseworthy College. An enquiry to the Mugford family confirmed that their eldest son, Keith, would indeed be interested in coming to work with us.

So it was that, in the early months of 1979, Keith left Adelaide and came to join us in the remote South-Western corner of Australia. For a time, he lived with us in our Busselton home. Sandra's recollection of that time is that it was "rather like gaining another son".

Our trusty retired shovel which helped plant many vines at Moss Wood, Smithbrook and Picardy.

I often reflect on the irony involved in the son of a former McLaren Vale general practitioner coming to work with us in Margaret River, on the recommendation of a member of the Hardy family, whilst my own eldest son, Stephen, has worked for BRL Hardy in McLaren Vale!

I am sure Keith won't mind me saying that, in his early days at Moss Wood, like most young men fresh out of tertiary educational institutions, he was a little green! What he lacked in experience, however, was more than compensated for by his great enthusiasm and very robust sense of humour. Keith and I formed a very strong friendship during his early months at Moss Wood and it has survived to this day.

Shortly after Keith's arrival, we carried out a retrospective tasting of all the wines which had been made at Moss Wood and agreed that the 1975 Cabernet represented the style of wine which we would both like to make. Accordingly, in the 1979 vintage, we picked the Cabernet a little less ripe than in the previous three years and at the same level of maturity as the 1975 vintage.

The resulting wine showed great refinement and balance. Although consumers seemed to prefer the style of the 1976 Cabernet, it has not stood the test of time as has the 1979, which is still drinking well today; the 1976 being well past its best. The 1980 vintage was what we would refer to as a "natural". A large crop of superb quality fruit ripened under ideal conditions. There is a saying in the wine industry that if you start with perfect fruit in the vineyard, all that you need to do in the winery is to avoid "stuffing it up". Such was certainly the case in 1980 and fortunately, as winemakers, we didn't stuff it up! The wines had brilliant colour and depth of flavour, together with superb balance. The 1980 Cabernet subsequently won the Montgomery Trophy at the 1982 Adelaide Wine Show and a reserve bottling thereof (which spent two years rather than the usual one year in wood) had become a benchmark for Margaret River Cabernet.

In 1980, we also made a small quantity of Chardonnay. It was a wine of considerable depth and richness which was a pointer to the region's future as a producer of great Chardonnay.

Then 1991 produced another near disaster. We were becoming rather concerned by the prevalence of powdery mildew (odium) in surrounding vineyards and had decided to use one of the new systemic fungicides on our vines. We now know that these should be used in combination with sulphur as the primary means of control. In our ignorance, we used the systemic agent on its own and were rewarded with a devastating outbreak of powdery mildew in the vineyard. What we managed to salvage from the Cabernet crop made a wine which, in my opinion, was just acceptable quality-wise. The saving grace of the vintage was a very small crop of Pinot Noir, which produced a wine which James Halliday has generously described as "Olympian".

It was at about this time that I underwent yet another "defining experience". Quite by chance, I had purchased a Bouchard Pere Chambertin Clos de Beze 1966. Sandra, Keith and I shared it over a late supper after a strenuous day in the winery. I must say that I had never tasted anything quite like it before (and rarely since). The intensity, complexity and finesse of that wine made an indelible impression on me which was to change my relationship to the vine and its products in a most profound manner.

It is also amusing to recount how a glass of that particular wine was offered to a South Australian wine judge, who happened to be dining in the same restaurant at the time. He declined the offer saying that he never drank "that rubbish" (Pinot Noir). Not many months thereafter, I happened to be reading a nationally distributed wine magazine which featured a Pinot Noir tasting - chaired by that same wine judge!

In 1982, Margaret River suffered the latest frost in its history as a wine producing area. It wiped out all of our Chardonnay and about half of the Semillon. A large part of our oldest Cabernet block was also affected. From what was left, however, we managed to make some pretty good wines. In the following year, at the Perth Wine Show, we won trophies for the best West

The vines at Moss Wood

Australian red wine and best one year old red wine in all classes with the 1982 Cabernet. The 1983 Semillon won the trophy for the best West Australian white wine. It was, for us, the realisation of a long held dream. Our joy was rather dampened, however, when the wine journalist for The West Australian failed to report our success in that newspaper. One can only speculate as to his reasons for this omission, but it became a bit of a joke among the wine cognoscenti that we had been "edited out". It was, for us, rather embarrassing when many of our friends enquired as to why we had not entered our wines in the Perth Show that year!

The 1983 vintage was a re-run of 1980, with a large crop of superb quality fruit which produced Cabernet wines exhibiting a depth of flavour which is only attainable with mature vines. The wine has required many years of bottle age to show its true class. I am confident that, in years to come, it will be seen as one of the greatest Cabernets ever produced in Margaret River.

In 1982, we had visited Burgundy for the second time. In the cellars of Comte Lafon, in Meursault, we had witnessed, for the

Keith Mugford: One of Australia most respected winemakers had his start at Moss Wood, where he remains as owner today.

first time, the wondrous complexity of Chardonnay which had been barrel fermented and aged on its yeast lees.

The custom in the Australian wine industry was, at that time, to "clean up" Chardonnay by filtration after it had been fermented in a stainless steel tank, before putting the wine into barrel. The resulting wines never demonstrated the perfect integration of fruit and oak, together with complex lees autolysis character which the Burgundians achieve with their Chardonnays. We determined that, on our return to Australia, we would use Burgundian techniques to produce our Chardonnay. We did so in 1983 and the resulting wine, when entered into the Perth Show scored nine out of a possible total of twenty points; suggesting a wine of truly execrable quality. I still have a few bottles of the 1983 Chardonnay and it is still a lovely drink, showing complexity and depth rarely seen in other Australian Chardonnays of that era.

Nowadays, it is rare to find an Australian Chardonnay of any class which is not barrel fermented and lees aged. Such is the price of being a trend-setter!

CHAPTER 5

ALL NOT WELL IN THE GARDEN

By late 1983, I was becoming more than a little disenchanted with my life in Busselton and Margaret River. Sandra and I agreed that the prospect of spending the rest of our lives in that community was one that neither of us found to be the least bit attractive. A joint expedition to the local supermarket would always result in discussion involving someone's bunions or haemorrhoids. When I gave a young mother and her child, who lived quite close to our home, a lift on a Sunday morning, someone immediately phoned Sandra to tell her that I had another woman in the car!

Disenchantment with Margaret River started to set in

Anonymity certainly has its virtues and the local doctor is far from being anonymous in a country town.

Besides which, there was a certain attitude in town which I used to summarize as "if the beer is cold and the fish are biting, life is just great!". Perhaps it was the Jesuit influence, with its heavy emphasis on the pursuit of excellence, which made me so unsettled.

I was becoming rather fatigued by the constant stress of treating victims of motor vehicle accidents and managing difficult obstetric emergencies. In urban practice, such cases can be referred to those more qualified and experienced in handling them. Subsequently, the stress levels in city practice are substantially less than in the country.

Our older children, who had received an excellent education in the nearby town of Bunbury, were at an age where they would shortly have to go to Perth for their tertiary education. We were therefore faced with the prospect of breaking up our family unit or moving to Perth to keep it intact. But there were other matters relating to the wine industry which I found quite upsetting.

Perhaps I should explain. In the early seventies, we set up a regional grape growers and winemaker's organization, of which I was the inaugural President. One or more of the frequent complaints which we received from visitors to the region was that they had great difficulty in finding many of the vineyards; being as they were, few in number and widely dispersed. During a visit to the Hunter Valley, Sandra and I noted that the local growers employed a system whereby a metal picket was placed on the verge at each intersection, and growers and winemakers would attach small metal signs to the pickets directing interested parties to their establishments. This system was simple, cheap and effective, and (one was silly enough to believe), the obvious solution to our problem in Margaret River. When I first raised the matter at a grower's meeting however, I was told that the proposal was unacceptable as it would lead tourists being directed to Moss Wood before any other vineyard! (At that stage, we were actually the most northerly vineyard in the region).

Vines in Willyabrup, Margaret River

I tried again, several months later, but the proposal was again rejected for the identical reason.

In the early eighties, Sandra and I made the first of many visits to the Champagne region in France. The Champenois are superb promoters of their product and have an unwritten rule that no producer of Champagne will in any way denigrate a fellow producer to a visitor to the region. Thus, when Bollinger is asked about Krug, or vice versa, the response is invariably "They make excellent wine". Their regional body, the Committee Inter-Professionel du Vin de Champagne, is highly organized and efficient in directing visitors to as many champagne houses as they can encompass during the course of their stay. One eventually leaves the region having been bathed in hospitality and a general air of bonhomie, feeling "What a wonderful area is Champagne". Of course, the Champagne region has its share of petty squabbles and enmities as does any other wine producing area. The great difference is that they are never exposed to the visitor. Added to which, the Champenois are extremely even handed in their promotion of each and every grower in the region.

Jancis Robinson OBE, MW, visited Picardy in 2001

Now it seemed to me that with the very small volumes of wines being produced in Margaret River at the time, and in view of the fact that the region was some four hours drive from Perth, (and who would drive that far just to see one or two wineries), that the area's producers had nothing to lose in emulating the Champenois in their approach to its promotion. One would have expected that referral of visitors to other wineries, particularly those that had produced excellent wines, would have been normal practice. Sad to say, this was most definitely not the case, and visitors to the region tended to be jealously monopolized by one winery or another.

During one of our visits to Burgundy, I spent a most enjoyable week acting as chauffeur for the British Master of Wine, Jancis Robinson, who was at the time preparing a documentary on the region. Not very much later, Jancis visited Margaret River. Much as I would have liked to show her over Moss Wood, we were not included on her visitor's list!

Whilst in Burgundy, Sandra and I visited a wine tasting centre that had been set up in an old church in the "wine capital" of the region, Beaune. The format was elegant in its simplicity and involved the purchase of a tasting glass and a directory, (with ample space for tasting notes) on entry. The tasting consisted of several stations, each with an opened bottle of wine on an upturned barrel. Each station represented a sub-region of Burgundy, (Gevrey, Vosne Romannee, Corton, Volnay, etc). A short explanatory text, and a representative wine was available at each. Sandra and I enjoyed the experience to the extent that we returned and repeated the tasting the following day. The concept was so worthwhile that we felt we really should attempt to set up a similar tasting centre in Margaret River.

Shortly after our return to Busselton, a neighbour, who was also a Shire Councillor, approached me with the news that Prospect Villa (one of the oldest and most attractive buildings in Busselton, adjacent to a BP service station and owned by the BP company) had been offered to the Shire to use as it saw fit. My neighbour wondered if I had any ideas as to a suitable use for it.

Now there are times in one's life when one feels that the Creator really wants something to happen and this was one of those times! The villa was situated in the centre of Busselton, with ample parking and opposite a large park where children could safely play while their parents attended to other matters. In short, here was the ideal venue for our tasting centre.

I reasoned that the centre would present a substantial benefit to small producers as it would obviate the requirement for them to provide tasting facilities and employ staff; thus reducing the need for them to be distracted from their viticultural and wine making activities. It seemed to me that if somebody enjoyed a particular wine, they could be provided with details of the producer together with the order forms and other relevant information. Thus there appeared to be no downside to being involved in a tasting centre.

Bursting with enthusiasm, I began, one by one, to contact the various producers in the region to outline the proposal to them. To my utter dismay, the proposal was flatly rejected by each and every one of them; usually on the grounds that it did not conform to "our cellar door sales policy."

I found it difficult to comprehend then, and still find it incredible, that as a group the Margaret River producers could be so absolutely unimaginative and egocentric in their attitude to the promotion of the region and its products!

At one stage, one of the region's winemakers was attempting to develop a system whereby genuine Margaret River wines would have a "Certificate of Origin" stamp on their label. At the same time, he was importing grapes from another region and selling the resulting wine under a Margaret River label! I questioned him as to how he could reconcile his proposal with his current activities. He responded that genuine Margaret River wines would carry the Certificate of Origin stamp and the rest labelled simply "Margaret River". I told him that he reminded me of the girl who felt that she might be "a wee bit pregnant".

The last straw came for me when in 1983 the Busselton Shire called for a submission with respect to a draft policy relating to ownership of food outlets and restaurants in the vineyard area. The draft policy, when released, had obviously been heavily influenced by one particular producer who served rather basic meals at his cellar door. It proposed that ownership of food outlets and restaurants in the vineyard area should be exclusive to wine producers. This was obviously a nonsense and I wrote to the council a most cordial and considered letter, in which I stated that in our travels throughout Europe, we had never encountered a single individual who had achieved excellence in areas of both wine and food production. The very obvious reason is that both disciplines are totally demanding and one will never make a great success of both. The Council neither acknowledged nor responded to my letter.

All of us have (at times) a need to feel valued. I believed then, as I still believe, that I had made a major and significant contribution to the Margaret River wine industry. To be treated with such contempt by the Busselton Shire was, for me, intolerable.

I decided that Margaret River and I had had just about enough of each other. I recall a particularly relevant passage from the Gospel of St Luke in which Christ tells his disciples: "Wherever they deny you a welcome, as you leave the city, shake off the dust from your feet". Busselton, even now, is hardly a city, but the above passage pretty much summarizes my feelings towards it.

I discussed with Keith Mugford the possibility of his buying Moss Wood. I had enough invested in the place (from the emotional point of view) to want to see it carried on as a distinct entity rather than just another marketing outlet for a large company.

Keith readily agreed to the proposition and in early 1986, when I was able to find a suitable general practice in a Perth suburb, Sandra and I moved our family back to the city of our births.

CHAPTER 6

BURGUNDY

After selling Moss Wood, we could afford to build a comfortable home in the Perth suburb in which I grew up. In addition we had some funds in reserve. We were comfortably well off but by no means wealthy and I had to keep working in my medical practice in order to meet the expenses associated with raising and educating four children.

Contrary to popular perceptions, wine making is not a very lucrative occupation and there is a saying in the wine industry that if one wishes to make a small fortune, one should start with a big one and buy a vineyard!

Be that as it may, we really enjoyed our return to Perth, catching up with old friends and revisiting old haunts. I well recall how excited we were on our first night back in Perth (in rented accommodation) when we heard a train go by on a nearby railway: It really made us feel that we were back in the civilized word!

The delightful Domaine de la Pousse D'or, Volnay, Burgundy

I have to say that I felt much more comfortable (from the intellectual point of view) among my friends in Perth than I ever did in Busselton.

The reader will recall that in previous chapters I referred to what I have described as "defining events" in my association with the world of wine. The first involved exposure to an early Grange Hermitage and the second occasioned by drinking a superb red Burgundy. One's tastes in wine do certainly develop with the passage of time and although, in my early days I was greatly attracted to the richness of the Grange, it was the combination of power and finesse in the Burgundy which really left me speechless. In the ultimate, the art of the winemaker is to layer richness and complexity on to a background of refined texture so that the wine, despite its obvious power, will retain considerable finesse. One never tires of drinking such a wine whereas one which is rich and powerful but lacks elegance will rapidly pall the senses.

After many visits, the charm of Burgundy never fades

Daniel and I tasting reds at Domaine Robert Chevillon

Many visits to Burgundy: Daniel and I with Frederic Mugnier

Daniel at Domaine Rapet tasting and discussing the wines

Thus, I determined that Shiraz and Cabernet could give great satisfaction but that ecstasy was to be found in great Pinot Noir.

From the early eighties, Sandra and I began to make regular visits to France and in particular, Burgundy. Frankly, I fell in love with the place. At home I was generally considered a bit of a bore as I tended to talk about wine constantly. In Burgundy, nobody talked about anything else!

The Burgundian winemaker is very "artisanal" in his approach to his craft. He tends his vines during the day and usually performs his cellar work in the evenings after long days in the vineyard. At first meeting, he can come across as being rather gruff and unapproachable. Having established, however, that one is seriously interested in wine and even more seriously interested in Pinot Noir and Chardonnay, there is no more affable nor amiable individual in the entire world of wine. On numerous occasions, we have commenced tastings in the early evening (after the vigneron has finished his work in the vineyard) and by the wee small hours of the next morning will still be sampling long forgotten vintages; bottles of which will be retrieved from obscure corners of musty cellars.

Breaking into Burgundy can be difficult. One can walk through a Burgundian village and see nothing but brass name-plates on doors. Only with direct introductions will the doors be opened to you. In this respect, we owe an enormous debt of gratitude to Becky Wassermann.

In 1981, we were directed to Becky by her then partner, Christopher Cannan of Bordeaux. She had originally left the United States to live in Burgundy with her artist husband Bert. Together, they had renovated and converted the barn of an old monastery at Bouilland in the hills above Beaune, making for themselves a most comfortable home in idyllic surroundings. While Bert spent his days painting, Becky began to search for something to do with her ample spare time. Initially, she exported barrels from Burgundian coopers back to winemakers in the USA. As time passed, she got to know many of the local winemakers and before long had an intimate knowledge of the Burgundian wine trade. Soon, she was exporting wine from small Burgundian producers to the USA and her home became a Mecca for English speaking visitors to the region. The "glitterati" of the wine world such as Clive Coates, Michael Broadbent, Robert Parker and Jancis Robinson were regular visitors and guests at her table. Becky became an unofficial directory for the wine tourist in Burgundy. All one had to do was to tag on to one of Becky's tour groups (more often than not as chauffeur or translator) and all doors were opened!

In those early years, I tried to develop an understanding of what constitutes a great Burgundy (particularly red) and what it should taste like in its youth, in barrel. This is not as simple as it sounds as great years in Burgundy occur on average, twice in each decade. Thus, in the time I have been acquainted with Burgundy, only 1978, 1985, 1990, 2000 and 2005 have been considered great years. In lesser years, there are, however, exceptions to the general rule: growers such as Henri Jayer of Vosne Romanee, who makes wine from very old vines, even in prolific years will produce quite small crops. Thus in 1982, when the vintage as a whole was characterised by rather thin wines produced from over-cropped vines, Jayer made excellent wines. Another who had a reputation for making good wine in ordinary years was Gerard Potel of Volnay, who pruned his vines to keep yields very low. Herein, I believe, lies the secret of producing great Pinot Noir: the nexus between low yields and high quality is inviolable.

The Burgundians have a saying, that with respect to Pinot Noir "finesse is everything", ("toute en finesse"). When tasting great Pinot Noir from barrel, one is not assailed by rich, fruity flavours, but rather by a sensation in the mouth of something extra-ordinarily complex but with a slight, green, tannic edge to it: something akin to eating fresh red currants. Interestingly, the higher up the scale of quality on greatness one progresses, from village to premier cru and on to grand cru, the more difficult are the young wines to assess. Grand cru wines of a great year seem to be so "tightly constructed", dense and impenetrable and will only begin to develop their wondrous opulence and flavour after many years in bottle. Ultimately, one has simply to accept the assurance of the winemaker that the wines from such great appellations as Richebourg, Musigny and Chambertin will, in time, show their majesty. This, of course, assumes that the grower in question is, by reputation, honest and competent.

Daniel and I with Denis Bachelet at
Domaine Bachelet, Gevrey Chambertin

Burgundy in the winter

Another fine Burgundian cellar at Domaine Tardy
with Daniel and Guillaume and John Tardy

Inspecting the low-trellised Grand Cru vines in Chabils

During the early eighties, Sandra and I spent many weeks tasting up and down the Cote D'or, from early morning to late at night, often returning to Becky's house to share some incredible old wines over the evening meal. Sandra often protested at the early starts but once we were on track she was, I believe, as enthusiastic as myself. The occasional offer for her to forgo a day's tasting was invariably declined!

On each visit to Burgundy, we would taste in excess of five hundred young wines (both red and white). As a result, I believe I have imprinted on my mind a very clear picture of what an excellent young Pinot or Chardonnay should taste like if it is to age with grace and balance.

As I have said, the best young, Burgundian Pinots tend to be very tight and rather lean in structure. With many years in bottle, they can develop a richness and opulence which is quite staggering when tasted. On the other hand, many "new-world" Pinots will have more fruit and flesh when young but will become fat, over-blown and unattractive after a few years in the bottle.

In the course of our visits to Burgundy, I had the good fortune, on several occasions to taste at Domaine de la Pousse D'or at Volnay. This was, for Burgundy, quite a large operation, with thirteen hectares of top quality vineyard producing about five thousand cases of most excellent red wine. The winemaker, Gerard Potel, was about ten years my senior but was, during the course of our visits, exceedingly warm and generous to us. He also seemed to be interested in our involvement in the Australian wine industry.

During the course of a dinner with Michael Broadbent, at Becky's house, in 1985, it was mentioned that a fifty per cent share in a great Burgundian Domaine was available if anyone was interested. The domaine in question was none other than Pousse D'or! The Seysses family held a fifty per cent holding in Pousse D'or, in addition to their own Domaine Dujac (at Morey St. Denis). It was their interest in Pousse D'or which had become available.

Some months previously, we had been visited in Margaret River by John Horgan, who had helped his brother Dennis set up the vineyard in the early days of Leeuwin Estate. I told John that I was thinking of leaving Margaret River and he asked me what I would be doing with my time after leaving Moss Wood. I responded that I would like to spend more time in Burgundy in order to learn more about growing and making of great Pinot Noir. I also told him that I would like, if possible, to procure a small piece of a Burgundian vineyard. John commented that if I ever found that small piece of Burgundy, he too would like to be involved. Obviously, the available share in Pousse D'or was very much more than Sandra and I could afford. On the off chance that he might be interested, I contacted Horgan and informed him of its availability. Horgan observed that the amount of money involved would be rather beyond his capacity also, but suggested that he approach several of his high profile friends in the finance industry with the proposal. Eventually a consortium of seven Australians (including Horgan and myself) was formed and in 1986 we purchased fifty per cent of Domaine de la Pousse D'or, the finance being organized through a Sydney merchant bank. The Potels were delighted with the arrangement as it allowed them to carry on running the domaine with minimal outside interference.

From my own point of view, our involvement provided me with unfettered access to the knowledge and skills involved in making great Burgundian Pinot. It fairly soon became evident that the majority of input and influence occurred in the vineyard (rather than the winery) and I tended to spend many hours working with the vineyard staff, discussing with them the underlying reasons for their various viticultural practices.

At the interpersonal level, our two families hit it off remarkably well. Sandra and Gerard's delightful and flamboyant wife, Francoise, instantly took to each other like sisters. In Gerard, I felt I had found a soul brother. With our common, intense interest in Pinot Noir, we would spend many hours discussing the subtleties and nuances of this most demanding grape and its wines.

Over the next several years, on more than one occasion, we swapped children: the Potel's only son Nicholas coming to live with us and being involved in several vintages in Australia while our own eldest son Stephen did vintages at Pousse D'or and at the famous Meursault Cellars of Comte Lafon.

During one visit we paid to the Lafon cellars, we were joined by two rather prissy English Masters of Wine. Not wishing to be intrusive, Stephen and I stood back and allowed the MW's to get on with their tasting. Suddenly one of them demanded of me, "What is your involvement in the world of wine?" I responded that I had owned a vineyard in Australia. "Whereabouts in Australia?", he demanded. "In Margaret River", I replied. "I suppose it wasn't something wonderful like Moss Wood", he proffered in his most supercilious tone. "That's the one!" I replied.

The reader will recall that the late 1980's were, from the financial point of view, very troubled times indeed! In particular, interest rates rose to levels which, even in those times, were considered horrendous. Our loans at the merchant bank were substantial and I found that I was having to work full-time in my medical practice, simply to meet the interest payments. Worst of all, we weren't able to spare the necessary time or money to visit our beloved Burgundy. How ironic that the acquisition which gave us the opportunity to participate, in a very intimate manner, in the workings of a Burgundian estate was also, for financial reasons, the cause of our being unable to take full advantage of that opportunity.

Be that as it may, in March 1988 Gerard and Francoise together with their friends Michel and Jacqueline Juillot (owners of a very large estate in Mercurey) visited us in order to observe, at first hand, an Australian vintage. During our travels throughout South-Western Australia, I took them to an abandoned wine grape trial-plot which had some years previously, been established by the Western Australian Department of Agriculture on their (then) research station between Manjimup and Pemberton. Gerard and Michel were obviously highly impressed by the quality of Pinot Noir grapes on the vines in that trial plot. During the course of our return journey to Perth Gerard and I discussed the possibility of establishing a vineyard in the Manjimup-Pemberton region.

Burgundy remains as charming as ever

CHAPTER 7

SMITHBROOK

Following the return of the Potels and the Juillots to Burgundy, I conducted a study as to the availability of suitable properties (for wine-grape production) in the Manjimup/Pemberton region. As luck would have it, a large property just to the north of the abandoned Manjimup Research Station (the site of the previously mentioned wine-grape trial plot), had recently become available. Inspection of the property showed it to be composed of beautifully drained, gravelly-loam soil, situated to either side of a ridge running from north to south across its centre. The property had good elevation and air drainage and would therefore, have very little susceptibility to frost. The only negative aspect was that the land had recently been used for vegetable production and the soils were heavily infested with nematodes. This problem could be overcome without great difficulty by grafting on to nematode resistant root-stocks. Above all, the price being asked for the land was extremely reasonable.

I proposed to the other Pousse D'or partners that we purchase the property and develop a vineyard and winery thereon, with myself, the Potels and the Juillots having major input with respect to the required knowledge and skills. Eventually a partnership was formed involving myself, the Potels, the Juillots and four of the other six Pousse D'or syndicate (or should I say, the private companies of those individuals).

For the next several years, I threw myself unreservedly into the development of Smithbrook (as the property became known), sourcing and directly collecting planting material, ridding the property of rocks and the massive karri tree stumps. Over the next several years, we planted some forty acres of vines, all of which involved a great deal of commitment and a lot of hard physical work on my part. This reduced my involvement and earning capacity in my medical practice to a considerable extent and we actually wore out a car on our trips to and from Smithbrook.

The question of any remuneration for all my input did not arise during the first two years of Smithbrook's development and I rather assumed that I would earn a share in the venture in lieu of payment for my efforts (I did, in fact, receive a return air fare to Burgundy which I used largely to research the availability and price of wine-making equipment, barrels, etc).

In the vineyard in 1991, we had succeeded in producing a tonne of Pinot Noir which I arranged to have processed by Keith Mugford at Moss Wood. When I first tasted the resulting wine out of barrel, I knew we were on a winner, as it had all the complexity and finesse of Burgundian Pinot. Before long, however, it became evident that problems were developing.

At the behest of the Australian partners in the venture (other than myself) the major proportion of the financing was by way of borrowings from the same merchant bank which had funded the Pousse D'or purchase. I had envisaged that we would all contribute the same capital to the venture. However, this did not eventuate and the project was largely debt funded.

The debt was structured so that bank fees and interest payments were "capitalized" (or added to the original loan) and fairly soon, the debt was, from my point of view, assuming alarming proportions.

I also realised that Sandra and I were borrowing from the bank in our own names, the other partners were operating through companies and I began to feel most concerned with respect to my level of personal exposure.

In addition, one of the Smithbrook partners announced, in 1989, his intention of setting up his own vineyard in the Pemberton region. Frankly, I considered this to be a most disappointing development.

In view of the various aspects of the situation which made me so uncomfortable, I put certain requests to the other Australian partners at a meeting in 1991. I requested that all partners consider the contribution of capital to the venture in order to lower the debt and reduce the interest burden. I also requested that I receive some financial reward for my substantial

Tall karri juxtaposed with vineyards has become a common sight at Pemberton

contribution to the project. My request for capital injection to the project was not accepted.

On the second matter, one of the partners was given the task of coming up with a suitable remuneration package for myself. When after several further months, the package had failed to materialize and the debt on the project was (in my view) still growing at an alarming rate, I resigned as project manager and as a director of the venture.

The next twelve months I can only describe as "hell on earth". My exit from the Smithbrook project was negotiated over that period and for Sandra and I this was a most difficult and traumatic experience: from both the emotional and commercial points of view.

Eventually, I was able to extricate myself from the venture and I suppose that there was a positive outcome in that I am now considerably less naïve with respect to business relationships.

Thus ended for us, the most painful and regrettable chapter in our involvement with the world of wine. The Almighty does, however, work in strange ways and the closing of the Smithbrook door opened another which set us on a course for a more interesting and exciting future. Most importantly, we were now in control of our own destiny!

CHAPTER 8

PICARDY

Traumatised and penniless as we were, Sandra and I took stock of our situation. On the positive side, we still had our health and our home in Perth. We had also, in 1989, managed to obtain a forty acre property near Pemberton by purchasing a whole farm and selling off the majority of it to interested parties, keeping what we judged to be the best of it ourselves. (This was our one and only foray into the entrepreneurial side of vineyard development.) Interestingly, when we had made preliminary enquiries about the property, we had been told that it was too stony to grow anything! (Didn't they used to say that about the Cote D'or in Burgundy?) Inspection revealed, however, the presence of superbly drained, deep, gravelly soils, ideally suited to the production of high quality table wine.

At that time, I said to Sandra that the aspect of my life which had given me the greatest pleasure was the growing of grapes and the making of wine and that it was my ambition to die with a pair of secateurs in hand!

I must say that over the most difficult period of my life, I gained considerable inspiration from John Kosovich, who is a highly respected winemaker and valued friend. John, who is some ten years my senior, had been developing his own vineyard at Middlesex (south of Manjimup) while I was involved with Smithbrook. I felt at the time that if John could take on a new vineyard at his age, then why shouldn't I at mine?

We had, as I had said, no money but I still had my medical practice and Sandra was working for an oil company in Perth. We were very hurt and angry about the way we had been treated by our ex-partners but reasoned that if we could convert our anger into positive energy and do most of the work ourselves, it would be possible to develop our own vineyard at Pemberton.

It would obviously be impossible for us to maintain our commitment to Pousse D'or and develop our own vineyard at the same time. I spoke to a friend and fellow Burgundy enthusiast, who had indicated an interest in acquiring a share in the Domaine

and he agreed to take it over. I then wrote perhaps the most difficult letter I have ever had to write and informed the Potels of our decision, together with an explanation of the underlying reasons.

The Potel's response was everything that I hoped it might be: that our withdrawal from the syndicate would not affect our personal relationship and that, as far as they were concerned, we were still "family".

We then discussed the possibility of developing a new vineyard and winery with our second son Daniel, who had recently graduated from Roseworthy College and was managing Rosa Brook Estate in Margaret River. Daniel agreed on the rather reasonable basis that he could, in the longer term, be guaranteed a significant interest in the venture.

At that juncture, I gave considerable effort to concisely defining our objectives in developing the new project. Clearly it was going to be our "last bite at the cherry" so that we would ensure that this one would be as close to perfection as possible. Consequently, everything that went into both the vineyard and winery would have to be of the very highest standard.

During my time in Burgundy, I had become aware of the existence of a very large clonal selection trial which had been carried on since the early seventies. In excess of two hundred clones of Pinot Noir and a similar number (of clones) of Chardonnay, which had been selected by growers as their best, were collected and grown in trial plots. The grapes from these plots were harvested and vinified separately. After maturation in bottle, the wines from each clone were submitted to blind, comparative tastings and the best clones were chosen, on the

basis that they produced the highest quality wines. (In other clonal selection trials, yields of fruit have tended to be more influential than wine quality). Initially, two clones of Pinot Noir (114 and 115) together with four clones of Chardonnay (76, 95, 96 and 277) were selected. A third clone of Pinot Noir (777) was later added. I had informed the partners in the Smithbrook venture of the existence of these clones but kept the knowledge as to how they could be sourced very much to myself (in view of the interest of the partner/competitor).

Pinot Noir was certainly my passion and we had made some quite reasonable wines from locally available clones (droopy/Pinot Tordu and upright/D5V12) at Moss Wood. The common criticism which the Bordelais make of their Burgundian cousins is that they make their wines from a single variety, whereas they themselves make their wines more complex by using a blend of varieties (Cabernet Sauvignon, Cabernet Franc,

Picardy: Our new beginning and my final shot at making great wines - Pinot in particular!

Merlot and Petit Verdot). This completely ignores the fact that there are over a thousand clones of Pinot Noir growing in Burgundy and that their growth habits, yields and flavours vary enormously. The message then, is to use as many clones as possible in order to give complexity. The "droopy" clone gives strawberry/melon flavours whereas the "upright" clone gives under-ripe cherry and currant flavours with more assertive tannins which add structure to the resulting wine. My Burgundian friends tell me that clones 114 and 115 add finesse with "classical Pinot flavours" to the wine whereas clone 777 gives mid palate flesh.

I was able to locate a source of both 114 and 115 in South Australia. We imported 115 in tissue culture (which involved changing the West Australian quarantine regulations) hoping to avoid having the material tied up in quarantine for twelve months. The tissue-cultured vines were, however, extremely small and fragile and didn't seem to give us any real advantage over importing cuttings through quarantine. We imported twelve cuttings of clone 114 through quarantine in 1994. The manager of the quarantine services glass-house, Malcolm Stewart (a careful, experienced and canny plantsman and quite a delightful human being) succeeded in producing about three metres of new growth from each cutting during its year in quarantine! Thus, at the end of that year, we had a total of some two hundred cuttings and plants of clone 114 Pinot Noir.

Clone 777 was a different story. I had actually been pursuing a source thereof since 1988 but could only ascertain that it was tied up in quarantine at the CSIRO station at Merbein in Victoria. For some obscure reason, the CSIRO officers at Merbein were unwilling to discuss the reasons for the delay in releasing the clone from quarantine. Through the good offices of a federal politician, I was able to direct an enquiry through the Federal Minister of Agriculture and ascertained that clone 777 was "partially" infected with a transmissible form of leaf-roll virus. Of the five vines which were originally imported, three appeared to be infected and it was only when sophisticated virus testing technology (ELISA) became available in about 1995, that CSIRO could declare two of the vines to be virus free and release them from quarantine.

I was promised Western Australia's entire allocation of five cuttings as nobody else had registered an interest in that particular clone. But there was a catch! The Western Australian branch of the Vine Improvement Association maintain monopoly rights to importation of any vine material from CSIRO into Western Australia. In this case, I wished to make a private importation (paying all the expenses myself) but, even so, had to apply to WAVIA for permission to do so. On receiving my application, they immediately registered an interest in clone 777 and took two of the five cuttings!

Malcolm Stewart again did an excellent job with my remaining three cuttings and at the end of their year in quarantine (1997-1998), I had the three original plants plus thirty one viable cuttings. These were planted out in the vineyard in the spring of 1998 and were given very special attention. At the end of the 1998-99 growing season, we were able to harvest enough material to give us a total of nearly four hundred vines! It should be said that not a scrap of such material is wasted and every match-stick sized piece is grown in pots before being transferred to the vineyard - usually in the glazed gallery of our Perth home, which makes an excellent glass-house!

Excellent soils were an encouraging sign

The situation with respect to Chardonnay was somewhat confusing. The original clone imported to Western Australia was "Mendoza". This clone is highly prone to setting problems ("chicken and hen" set with more chickens than hens) and seemed, from our experience to give a fatness and oiliness to the resulting wines. During a visit to Perth in 1989, the head of a very famous Burgundian white wine producer Domaine Leflaive (M. Vincent Leflaive) commented on the character which he strongly disliked in Australian Chardonnay was a "fatness" or "oiliness". What he looked for, but did not find, were "floral" characters on the nose and freshness on the palate. The other available clones of Chardonnay were Californian selections which seemed to produce enormous quantities of fruit with very little flavour. Being anxious to avoid "oiliness" in our Chardonnay, we eschewed Mendoza. When we examined the descriptive literature concerning Burgundian selections, we noted that "floral characters" were common to all of them.

We reasoned that, if we were to produce a Chardonnay in a country which was already awash with that variety, it would have to be something really exceptional. We determined to start again from "scratch" with the Burgundian clones. The available clones were 76, 95, 96 and 277, so twenty cuttings of each were imported through quarantine in 1994 and we have bred commercial quantities from that small beginning.

As part of the 1999-2000 season, we have over a thousand vines of each clone producing grapes. Of the clone 115 Pinot we have two thousand mature vines (half of them grafted on to low vigour root-stocks) and fifteen hundred vines of clone 114 Pinot.

Daniel's great passion is Shiraz. Not the fat, over-ripe and alcoholic, over wooded styles then being lauded by Australian wine-show judges, but the more subtle styles resembling those of the Rhone Valley in their texture and flavour. John Kosovich was kind enough to provide us with cuttings from vines which he (and his father before him) had progressively selected from the original Houghton plantings.

We reasoned that the market for Pinot Noir is far from universal and that not everybody likes (nor is able to drink) Shiraz, so we would need to produce a Bordeaux style wine. As with St Emilion/Pomerol, Pemberton would be (climatically speaking) rather marginal for Cabernet Sauvignon: not being warm enough to fully ripen the variety in many years.

We decided to follow the example of the "left-bank" producers of Bordeaux and plant a predominance of Merlot (fifty percent) with lesser amounts of Cabernet Franc and Cabernet Sauvignon (twenty-five percent of each).

Sandra and I personally collected all the Shiraz cuttings from the Kosovich's Westfield (Swan Valley) vineyard and the best clones of the Bordeaux varieties from Moss Wood (with Keith Mugford's blessing). All of this material was grown in a nursery at Picardy for the 1993-94 season and was planted out in the vineyard in the following year.

Planting the first tissue cultured 115 vines at Picardy, Pemberton

Sandra, myself and various friends did the majority of the setting out and planting with a contractor being engaged to prepare the ground and ram trellis posts. Sandra and I trained all the vines during our fortnightly visits to the property, often running strings from the small vines to the trellis wires by moonlight. On chilly, winter nights, with the magpies whistling in the trees and the mist rolling out of the nearby forest, it was actually a most exhilarating experience.

In 1996, we harvested some five tonnes of Pinot Noir from the original "droopy" and "upright" vines and this was vinified by Daniel, at Rosa Brook Estate.

Having made our first wine, we felt we had better find a name for the property. We began to compile a list of "possibles", scouring road maps of Australia, the United Kingdom and France for inspiration. From the earliest days at Moss Wood, we had enjoyed the close friendship and support of Patrick and Christine Stephenson, (the proprietors of the Liberty Liquors chain). Through the darkest days of the Smithbrook affair, they gave us unstinting support, even to the point of offering financial assistance if required; an offer which we declined in order not to prejudice a valued friendship. When discussing our quest for a name, Christine observed that her native region of France, Picardy, did not produce a wine of its own and that the name had a pleasant ring to it. Several years previously, we had stayed with the English wine merchant, Kit Stevens, at his home in Cognac. Kit's dictum was that all great wine names had to begin with "P", so that the Americans could pronounce them. When one considers Petrus, Pichon, Palmer, Pomerol, Pauillac, Pommard, it would seem that he has a point.

We showed our list of "possibles" to a large number of friends and acquaintances and the almost universal choice was - Picardy!

CHAPTER 9

NOT ANOTHER TIN SHED

My mother Nita Pannell in 1936

My father Jim in 1917

I have not, as yet, mentioned my parents.

My mother was a school teacher who had, in her later years, developed a passion for the stage, eventually gaining something of a reputation in that area and being honoured with an Order of Australia and an OBE for her achievements.

My father migrated to Australia when he was twelve years of age, leaving the green fields and rolling hills of Oxfordshire for the dry brown wheat-belt country of Wickepin. He took to this new environment with enthusiasm and loved the clear blue skies, the vivid colours of the local birds and the endless variety of fauna which surrounded him. Nevertheless, he was a rather introspective individual, being the only child of a rigid and angry father and a devoted mother.

After several years helping to clear and crop the new farm, interspersed with interludes of education activity (initially at the local bush school and later with the Christian Brothers in Perth), he was encouraged by his mother to study medicine. This required a move to Melbourne for some six years, after which he returned to Perth and met and married my mother. My elder brother and sister were born over the next several years. The family then moved to the small wheat-belt down of Goomalling where father purchased a general practice. The Great Depression struck shortly after their arrival in the town and they (along with all the other local residents) found times extremely hard.

Father's ambition had always been to become an orthopaedic surgeon and he desperately wanted to return to the United Kingdom to undertake further training and obtain his surgical degree. With the advent of the Great Depression, followed by the Second World War, however, it was not until 1947 that he was able to do so. Meanwhile the rest of the family stayed in Perth (or rather the nearby hills hamlet of Darlington). I recall this as a very happy time as we used to ride our horses to the local school and spent all our spare time exploring the local bush. Mother did join him in England for a year or so and I was put into a Catholic boarding school (not a happy time)!

When father returned from Britain, he was one of a handful of trained orthopaedic surgeons in Perth. As a consequence he was kept extremely busy, performing emergency surgery at all hours of the day and night at Royal Perth Hospital, largely on an honorary basis, as was custom in those days.

It was perhaps, as a consequence of all the above activities that father and I saw very little of each other in my early years and subsequently developed very little in the way of common interests or understanding.

Sandra and I with my parents Jim and Nita - and the dogs!

I often had the feeling that I was a bit of an afterthought and that by the time I came along, my parents were too busy with their various activities to have any major interest in my development. As a life-long friend says, I brought myself up. Whatever the reason, I cannot recall father actually congratulating me on any achievement. Indeed, in later years, if I tried to discuss matters relating to vineyards or wine with him, he would promptly change the subject. On the other hand he was more than pleased to discuss my activities at Moss Wood with his friends and distribute bottles of the wine I had made as presents.

Father had always had an active and avid involvement in rowing, (in his early years he lived quite close to Henley on Thames) and was himself a successful King's Cup oarsman. In his later life he developed a passionate love for Italian opera, (lord knows why, as he didn't seem to like Italians or any other aspect of their culture). Mealtimes could be quite bizarre with mother sitting at one end of the table, prattling on about theatre and father learnedly discoursing about rowing or opera at the other, with no common ground in between!

In the early 1950's, I was sent to the local Jesuit College for my education. The Jesuits, true to their reputation, were hard men, but fair. Their well known "modus operandi", "give me the boy and I will shape the man", is by no means an exaggeration. The commitment to excellence and capacity for self-assessment which they inculcate into their students (as often as not on the end of a piece of leather) once instilled, never leaves one. I suspect it is a major factor in my eternal and urgent quest to make the perfect wine.

Sadly, for the last years of her life, mother became demented. How painful it was to witness the rapid deterioration from one who could memorise whole plays with consummate ease to a little, old lady in a nursing home, barely able to recognise members of her own family. She died in 1995.

Young Clone 277 Chardonnay plantings

In father's later years, as my brother and sister were not living in Perth, a good deal of his day-to-day care fell to Sandra and myself. Despite this, he never ceased to treat me as a parent might treat a very young child. Sadly, we never at any time managed to develop an "adult-to-adult" relationship. In his final year, father developed gangrene and had to have a leg amputated. At ninety-five years and in one so proud of his physical prowess, this was a major disaster! As a medical practitioner, I can only wonder at the wisdom of keeping people alive long enough to inflict mutilating operations upon them.

Father eventually died, in his own residence, in 1996. As he died, being aware that hearing is one of the last senses to go, I kissed him and whispered to him what a pity it was that we had never truly understood each other.

Again, as with Smithbrook, I was determined that something truly worthwhile would emanate from our relationship. In his will, father left us sufficient funds to build a small winery and purchase the basic wine-making equipment. In my own heart (and in Sandra's also, I'm sure) we regard the winery as a monument to his final generosity and a legacy to his successors.

Sandra, for her part, was adamant that we would not build another "tin shed" such as we had at Moss Wood.

We had been closely observing the construction of a house in a nearby suburb. The materials employed were essentially processed cement "limestone-look-alike" blocks and French canal tiles and the overall effect was unusual but visually most pleasing. The blocks represented a very cost-effective method of construction, but French tiles? A call to the local agent revealed, surprisingly, that they cost very little more than locally produced roof-tiles.

We pored over every available tome on French rural architecture and determined that the Provencal "limestone and canal tiles" look would be just right for Pemberton. Daniel and I carefully analysed our spatial requirements for fermentation, tank, barrel and packaged product storage, based on projected production volumes from an eight hectare vineyard. On this basis we drew up a rough floor plan which was then refined by an architect. A builder was engaged and construction commenced in mid 1996.

Our eldest son, Stephen (by then a senior winemaker with BRL Hardy) was consulted regarding the selection of wine-making equipment. He had no hesitation in recommending a "Delta" crusher-destemmer, produced by the French firm Vaslin-Bucher. Fortunately, that firm also produces an extremely refined and efficient pneumatic press and a must-pump matched to the Delta crusher. We were able to do a "bulk-deal" with the local agent and acquire all of this basic equipment at a very reasonable cost.

The red wine fermenters were based on a design which I had observed at Domaine Roy in Auxey Duresses, Burgundy. They were essentially a closed tank but with a very large lid which allowed for hand plunging (pigeage) or even breaking up of the cap by the winemaker's legs and bodies in the traditional Burgundian manner. On the other hand, these tanks could be closed and the must capped with carbon dioxide, if we wished to carry out a pre-fermentation maceration without the risk of oxidation. The fermenters were placed on elevated stands so that the pomace could be dumped directly into the press through a door in the bottom of the tank, without the intervention of a must pump, thus avoiding undesirable maceration of seeds and stalks.

The fermenters and storage tanks were all produced by an Albany firm of tank makers. I can only describe the quality of their work as impeccable.

Daniel and his young wife, Jodie, moved to Picardy in late 1996, in time to supervise the completion of the winery and preparation for the 1997 vintage. Sandra and I had renovated the old house on the property and the young people were able to make it their home. I think it is fair to say that Sandra felt a little "displaced" at that time and it wasn't long before we were considering construction of a house for ourselves on the property.

The 1997 vintage was, from the climatic point of view, most successful and the winery and our new equipment functioned without a hitch. The 1996 Pinot Noir which Daniel had made at Rosa Brook was transferred to the Picardy winery, in barrel early in 1997. It was bottled on completion of the 1997 vintage.

We felt that the 1996 Pinot had quite delicate texture but

In goes another bucket load

Sandra with vines working in the vineyards during winter pruning

Three generations of Pannells in the vines: myself, Rhian and Daniel

Our first bottling of Pinot in 1996 meant we were up and running

Tables set in the winery for the opening of Picardy - a great occasion!

Picardy opening: Keith Mugford, Peter Forrestal, Anne Gregg, Sandra and John Leydon

Picking crews in the Picardy vines where we often have 50 working

Vintage 1998 saw the birth of my first grandson. Jodie on the Massy at eight and a half months pregnant

Grapes in buckets heading for the Picardy winery in 2011

lovely Burgundian flavour. Nevertheless we were unwilling to jeopardize our long-term reputation by releasing a wine which was not of at least acceptable standard, in the first instance. We showed it to a variety of friends and "amateurs" and even to several visiting Burgundian winemakers. They assured us that the wine was of a "respectable" standard and we released it for sale in October 1997.

Imagine our dismay when a Melbourne wine scribe, who had not even been asked to review the wine, described it in his newsletter as "green and stalky". For some weeks, we conducted a dialogue with him, saying that this was a wine from young vines, made in the Burgundian manner (for long-term development rather than early consumption) and only recently bottled. He was totally unrepentant, and told us (in a manner reminiscent of a first year undergraduate lecturing his professor) that we should not be marketing wine from such young vines!

I have a saying with respect to my beloved Australian rules football, that it is a very easy game to play from the comfort of the grand-stand. Planting vineyards and developing wineries is an extremely expensive past-time. If one is to maintain a policy, as we do, of only producing wine from grapes that we have grown ourselves, then one has to sell wine from young vines in order to survive financially. Besides which, first releases, though not great wines are an essential part of the history of a boutique winery.

Be that as it may, we were thrilled several months later, when we received the results of a local University Wine Society blind tasting of a range of selected Pinot Noirs from France, Australia, New Zealand and California. The 1996 Picardy came in just behind the top wine of the tasting (a Domaine de la Romanee Conti Echezaux) and well ahead of the rest of the field.

I sent a copy of the results to the afore-mentioned wine scribe, together with my best regards. Oddly enough, he has never responded on this matter!

No matter how assiduously one attempts to predict how a certain area will perform or what style of wine it will produce, the only certainty which exists is what the wine actually tastes like when it is ultimately put into bottle. It began to seem, as far as could be judged from our 1996 and 1997 Pinots, that the outstanding features of Pemberton red wines would be vibrant hues (rather than the deeper but duller reds of Margaret River), racy and delicate varietal flavours and extraordinarily low pH levels at moderate acidities. They also seemed to have exceptionally fine-grained tannins which (another) Melbourne wine scribe was kind enough to describe as "Bordeaux-like". All of this suggests wines which will need ageing to show at their best but which age gracefully for very long periods of time.

On a sad, sad note, we received a telephone call from Nicholas Potel on a stormy night in October 1997, to say that his father had died of a heart attack on the previous night. Sandra and I were devastated by the news. I felt as though I had lost a brother. I did consider going to Burgundy for the funeral but Nicholas and Becky Wasserman counselled against it, saying that our presence would be more valuable when the family had recovered from the impact of this tragedy. In retrospect, I wish I had gone as I still have a rather weird, empty feeling within me that I have never said goodbye to Gerard.

It had been my great ambition to share a bottle of our first wine from Picardy with my most esteemed friend and colleague. Alas, it was never to be!

The 1998 vintage at Picardy was one which the Chinese might well have described as "interesting". It was obvious, from fairly early on in the season, that the local marri trees were not going to flower and that the predatory small birds (silver eyes), would be a problem for us. Nevertheless, we had quite a large crop on the vines and were prepared to sacrifice some to nature and as Jack Mann used to say "the little birds sing so sweetly". What we did not anticipate was the seventy five millimetres of rain which fell over two weekends, just as the Shiraz and Cabernet were coming up to full ripeness. Fortunately, we had already picked the Pinot Noir and the Merlot prior to the downpour. The Cabernets didn't seem to be badly affected but the Shiraz rapidly picked up water and showed evidence of dilution of flavour. The rain was followed by several weeks of uninterrupted sunshine and we were able to get the Shiraz back up to 13 degrees Baume, but with some losses to the hordes of hungry birds.

Much to our surprise, the Cabernet Sauvignon soldiered on and continued to ripen despite the rain. Apart from minor bird damage, we eventually harvested quite a large volume of excellent quality fruit.

There was, however, a "production bonus" to the 1998 vintage. Our first grand-child Che David, was born to Jodie and Daniel on the 28th of April 1998. Like all good vignerons, he timed his entrance to perfection, delaying his arrival until just after the last grapes of the vintage had been harvested.

Tank sampling the Chardonnay

CHAPTER 10

STAVES OF OAK AND HOOPS OF STEEL

The wines of the 1997 vintage were well reviewed and well accepted in the market place. There was one aspect of their production with which we were not entirely happy.

For the 1996 vintage, we had purchased a number of barrels from a reputable French "tonnellerie". We were very happy with the quality of our initial purchase and ordered a larger number for the 1997 vintage. The oak in the second shipment was, we felt, a little under-matured and added a barely perceptible touch of "greenness" to the wines. This was not enough to be a serious problem but disappointing when one produces superior quality grapes and is trying to make wines of supreme quality.

Our philosophy at Picardy it to attempt to achieve the maximum possible level of control over every aspect of the wine production process. Here, clearly, was a problem which was a little beyond our control. With this in mind, I decided to spend a little time in Burgundy in an attempt to gain an understanding of both the basics and complexities of the oak barrel industry.

In October of 1988, armed with the details of several recommended barrel makers, we made our way to Burgundy and lodged in a "gite" next door to Becky Wassermans's home. It was marvellous to catch up with Becky and her husband Russell again after a break of some seven years. We also renewed acquaintances with Francoise and Nicholas Potel and found that little had, in fact, changed as far as our relationship was concerned. Nicholas was developing a new life for himself as a negociant; making wine from grapes he had purchased and buying, rearing and bottling wines which he had bought from other producers.

Francoise had purchased a small, comfortable house in a suburb of Beaune and was, as much as possible, surrounded by her children and grandchildren. On the anniversary of Gerard's death, we organized a picnic at the top of a small mountain overlooking the Cote D'or. Francoise seemed delighted to have

us with her at that particular time and one had the very strong feeling that Gerard was with us in spirit.

Our main objective, however, was to visit every known tonnellerie (cooperage) in the Burgundy region. I did have references to many of them but consulted the yellow pages in the local telephone directory to find still more. One of those was listed at an address in the town of Nuits St. George. When I called the relevant number, an elderly man answered and informed me that the building was no longer used as a tonnellerie. He did say, however, that a friend of his had recently started to produce barrels in a town quite close to where we were staying.

Having made an appointment, we visited the tonnellerie of Daniel Meyrieux et Fils at the village of Villars la Faye. We were immediately impressed by the level of care exercised in the selection and maturation of the oak staves. The manner in which the barrels were assembled and toasted was, we felt, exceptional. The toasting process causes slight charring of the inner surface of the barrel and will ultimately give flavours of vanilla, coconut and coffee to the contained wine. Normally this is achieved by placing a quite vigorous fire (in a metal pot) inside the upturned barrel and leaving it there for about twenty minutes. Meyrieux employ a rather unique process in which a low intensity fire is left inside the barrel for about ninety minutes. This causes less charring but a deeper toasting effect on the staves. The flavours which this imparts to the wine are quite subtle and eminently suited to the delicacy of Pinot Noir. What really impressed us was that Daniel Meyrieux took us to the cellars of several producers who used his barrels. The wines tasted absolutely delicious with subtle, supportive (rather than dominant) oak flavours. Both of these producers spoke very highly of both M. Meyrieux and his barrels. One of these was a producer whom I already knew well and greatly

The Picardy barrel room

admired: M. Dennis Bachelet of Gevrey Chambertin. The other was M. Jean Tardy of Vosne Romanee, who introduced us to his son, Guillaume, of whom more will be said in due course.

We also visited many other barrel makers whose product was generally of a very high standard. We had used many of their barrels at Picardy but most had existing arrangements with Australian importers and were unwilling to deal directly with us. We reasoned that if we could become agents for a particular maker, it would improve our chances of being supplied with a consistently high standard of cooperage.

To that point, Tonnellerie Meyrieux had worked only in the northern hemisphere (France and the USA) but were at the point of considering exploring markets in the southern hemisphere, in order to even out their workload. How fortuitous that we should arrive on the scene just at that particular time!

We therefore committed ourselves to purchasing fifty Meyrieux barrels for the 1999 vintage at Picardy; ordering a combination of Bourgogne, Nevers, and Vosges oaks with promises of a wider selection in future years.

Back at Picardy, our grandson's arrival made conditions in the existing house on the property rather crowded. It became obvious that our programme for a house for Sandra and myself would have to be brought forward.

We chose a design based on a typical "maison a bolet" from Quercy, in the south of France. The style of the house would suit the sloping site which we had chosen for it and would have an underground cellar. Moreover, the dwelling would be visually pleasing when viewed from any angle, an important consideration when constructing a dwelling in the middle of an open space. We chose rough, limestone blocks for the corners (coins) and window surrounds and an earthy tone of render to the exterior walls. Tiles to match those of the winery were selected for the roof.

Westerly view through the vines to the winery

Harvesting the Cabernet Sauvignon 2011

We were fortunate enough to find a local builder (of Italian extraction) who obviously had a "feel" for that type of construction and he did an excellent job. In view of the fact that we were building the house a year or so earlier than we had originally planned, we were a little financially constrained and did much of the internal finishing ourselves. To be frank, I quite enjoy tiling as I find it quite creative and not too difficult from the technical point of view. Painting is a "horse of a different colour" and I loathe it with a passion!

A simple garden, in keeping with the rustic design of the house, has been planted and a small vineyard, planted on identical spacings to those in Burgundy has been established nearby. For some time I have had an urge to "explore the envelope" to the ultimate degree with respect to Pinot Noir and my "toy vineyard" (as Daniel refers to it) is an essential part of that process.

Picardy winery in autumn

Several years previously, we had planted an olive grove between the house site and the road and this both screens us from passing traffic and adds an air of Provencal authenticity to the house.

The 1999 vintage occurred while the house was in its middle stages of construction. This caused Daniel some degree of annoyance as, despite our best efforts, he was often distracted from his vineyard activities by the demands of the builders.

The vintage looked anything but promising at the outset. A very large crop had to be significantly reduced by hand thinning of the bunches. Selected rows of Pinot Noir (encompassing all the different clones) had their crop reduced to five tones per hectare at veraison (colour change in the berries). Bunch exposure in those rows was improved by hand plucking all the leaves from the fruiting zone (a most arduous task).

From the beginning of February to the first week in March, the weather was cool, overcast and drizzly. It looked for all the world like a disaster year in Burgundy and we feared the worst! From the second week of March, the weather improved and we were blessed with clear, sunny days and maximum temperatures in the mid twenties right through until the end of May. Each variety, in succession, came slowly but steadily to perfect maturity with extraordinary development of flavour and perfect balance. Looking at the wines in barrel, I feel they are probably the best we have ever made. People look askance at me when I say this, as if I am indulging in a bit of pre-release promotion for the 1999 vintage. It is my honest opinion so what else can I say!

In 1999, we succeeded in making our first commercial quantity of Chardonnay from the new clones. The wine was barrel fermented and matured on its lees in casks ranging from new to three years old. I had mixed feelings about lees stirring in Chardonnay (batonnage) until I saw the same wine (a 1998 Corton Charlemagne) in two identical barrels, one stirred and the other not at Michel Juillot's cellars. The difference was truly remarkable: the stirred wine having much greater depth and complexity than the other. Needless to say we are now committed to stirring lees in our Chardonnays!

The Pinot Noir which was heavily crop-thinned, was

fermented separately and has produced a wine of exceptional quality - further evidence of the nexus between low yield and excellence.

This is probably as good a time as any to raise the matter of Australia's ability to perfectly ripen a range of varieties in the one location. In France, nobody would consider growing Shiraz in Burgundy. Michel Juillot actually tried and could never get it any riper than 9 degrees Baume! I believe the answer lies in the fact that in the European situation the transition from Autumn ripening conditions to low winter temperatures is generally quite sudden and dramatic. Conversely, in Australia, we tend to have a gentle descent of temperatures at the end of the ripening season and often an "Indian Summer". In other words the "window" available in which perfect ripening conditions persist is much longer in Australia than Europe.

The Meyrieux barrels more then lived up to our expectations and we will certainly be placing repeat orders in future years. Daniel Meyrieux and his partner Martine visited us in May 1999 and Daniel was most helpful in his observations regarding the matching of particular types of oak to different varieties. For instance, Bourgogne and Vosges seem suited to Pinot Noir and Chardonnay and Nevers more suited to the Bordeaux varieties. He also made some interesting suggestions with regard to subtle changes in toasting techniques for each variety.

The wines of the 1998 vintage surprised us a little by the way they "closed up" after bottling, probably on account of their extremely low pH levels. This transient loss of flavour, body and even colour can be quite disconcerting and was certainly so in this instance. Even the Merlot-Cabernet which had tasted quite fleshy and opulent in barrel looked quite thin, hard and uninteresting for several months after bottling. Fortunately, they had started to "open up" again by the time they were released in October 1999 and their acceptance by (and feedback from) restaurants, retail outlets and the wine press has been most positive.

At this point, I allow myself to dare to believe that we have almost made a success of our second vineyard and winery.

Fruits of hard labour

An aerial view looking south of the vineyards and winery at Picardy, Pemberton. The cooler southern region is unique and has a very promising furture for producing a variety of wine styles, particularly the classic Burgundian varieties for which Picardy has gained much recognition in our relatively short history.

CHAPTER 11

THE REASON WHY

One question which I have been asked more often than any other is "why is it always the doctors who plant vineyards?" This is usually followed up by a rather derogatory "I suppose it's because they have the money".

On the face of it, the question is probably not unreasonable. In the annals of Australian wine, the medical profession has been unusually well represented. Doctors Penfold, Lindeman and Angove are generally recognized as patriarchs of the industry. In Western Australia, Dr John Ferguson, a colonial surgeon, initially set up the Houghton vineyard. In Margaret River, Tom Cullity planted the first commercial vineyard, Vasse Felix, and was followed by myself at Moss Wood. Doctors Cullen and Lagan followed hard on our heels with Cullen and Chateau Xanadu respectively.

The proposition that wealth is a significant factor does not stand up to close scrutiny. The majority of the medical profession who have involved themselves in the wine industry have been relatively lowly paid general practitioners rather than their more

Our home amid the vines

well-to-do colleagues in the specialties. Unlike those from the business community, who tend to employ labour to perform vineyard and winery activities, the "wine doctors" do seem to enjoy a degree of personal involvement. In simple terms, they like to get their hands dirty!

What then, is the motivating force which specifically attracts medicos to the wine industry?

Tom Cullity used to assert that the two disciplines are quite similar in that for success to be achieved in either, infinite attention to detail is essential. Obviously, in medicine, lack of attention to detail will lead to mistakes which may well have tragic consequences and tend to be poorly tolerated by one's clientele! The multifaceted nature of the winemaking process and the multitude of factors which ultimately contribute to wine quality (soil, climate, clones, vine-nutrition, training and trellising techniques, flavour, sugar and acid levels, fermentation techniques, choice of yeast, selection of oak, etc) obviously demands great attention to detail if the process is to ultimately result in the production of high quality wine.

I have spent many hours, over many bottles of very good wine, discussing the phenomenon of "doctors and wine" with a psychiatrist friend. He makes the point that medical practice is almost entirely reactive. Our patients bring their problems to us and we react to them. On the other hand, the winemaking process is highly creative. My friend asserts that mankind has a great need for the creative and that if one spends the majority of one's time in reactive activity, this drastically increases that individual's requirement for creative activity.

When I apply his theory to myself, it certainly seems to fit. I simply love growing things, be it vines, olive trees, roses or vegetables. Having finished an exhausting session in the consulting room in Perth and then driven for four hours to Pemberton, there is nothing that pleases me more than to get out among my vines (with a flashlight) to see how they are progressing. When I do so, I feel as though a great weight has been lifted from my shoulders.

The aspect of viticulture which I most enjoy has mainly to do with developing young vines. To start with a small number of cuttings of a particular, imported clone and (over several years) to breed these up to the point where they number in thousands, I find extremely rewarding and satisfying. I never cease to marvel that a bud about half the size of a finger-nail can, within a month of being grafted on to a root-stock, produce a metre or so of healthy vine shoot.

I recently read of a shop in Perth which is doing a roaring trade in "relics" of a bygone era, such as blade razors, fountain pens, fob watches and snuff boxes. Is it possible that in today's homogenised, stereotyped and computer controlled world, some long for an element in their lives which is a little less than predictable or has a certain "hand-made" quality?

Wine, particularly at the boutique level, certainly fulfils both of those criteria. Its quality and character will largely be controlled by unpredictable, seasonal influences, resulting in the well known vintage-to-vintage variation. They exemplify the "hand-made" quality of boutique wines, relative to the stereotyped character of the large wine company product. The latter may be technically perfect but it lacks the "thumb-prints" of the maker.

As I have said, when great wine is made, one has the feeling of being an interested bystander, observing nature achieving perfection!

And then there is the joy of revisiting times long past, in which one participated in this minor miracle, simply by the drawing of a cork. In a way, wine is simply bottled history!

Then, for me, there is the Jesuit influence. The Jesuit credo is that our Creator made us with certain talents and capabilities and that which pleases Him most is that we should use those talents and capabilities to the maximum, in much the same way as a parent feels fulfilled when observing the accomplishments of a child. To a Jesuit, prayer is simply a matter of using every minute of the day in some positive and constructive manner. Mind you, this can have its downside. I find I am quite unable to enjoy a holiday sitting by a pool or looking at a view. I simply have to be doing something productive! On the other hand, wine is generally grown in the most delightful places which, more often than not, have their own unique culture and cuisine. Bordeaux seems to be a glaring exception as it neither has a distinctive cuisine nor, most unusually for France, its own cheese! Frankly, there is nothing I enjoy more than visiting winemakers and talking to them about what they do and why they do it.

What is more, when one is involved in the wine industry, the world tends to come to one's door in the form of visiting winemakers, journalists and even the "League of Nations" mob of back packers we usually employ at vintage. In essence, it could be said that the winemakers of the world form a vast fraternity who are more than willing to welcome other winemakers and openly exchange ideas.

On looking back over my life, I have to acknowledge that there have been times when it all seemed pointless and too difficult. Overall, however, I believe that I have (to use a common colloquialism) "had a go". I suppose that the ultimate test is to ask oneself: "If I died tonight, would I regret not having achieved more in any particular area of my life'? I believe I can honestly answer - no.

CHAPTER 12

INTO THE NEXT MILLENNIUM

Making great Pinot Noir is a passion that will never fade for me

At the end of a somewhat tortuous journey over the past thirty five years, we arrive at the present and the eve of the twenty first century.

What, then, is one's vision of the future?

The wine industry is, once again, facing a period of major instability. A quite irrational burst of vineyard expansion has seen targets for the year 2025 already achieved. Unfortunately this has not been matched by an increase in processing capacity and a massive over supply of grapes has been virtually guaranteed, in the not very distant future. In turn, this will precipitate an over supply of cheap, average quality wine as growers, anxious to sell their highly perishable crops, will accept prices which will, in many cases, be less than their costs of production. Hopefully, the majority of this wine will find buyers overseas. On the international market, average quality wine is a highly price sensitive commodity and the surplus of grapes will need to be purchased, by the wine companies, at extremely low prices if the resulting wine is to gain market share against countries with similar problems of over supply. Even so, an over supply situation must necessarily have a depressing effect on wine prices on the domestic market.

How will Picardy survive this vinous Armageddon?

The mantra of the marketing gurus is that every producer in the wine industry must have a "marketing plan". When asked to be more specific, there is usually much shuffling of feet and looking at the floor! To my mind our best marketing plan at Picardy, as it was at Moss Wood, is quite simply to make the best quality wine and to market it at affordable prices. Hopefully, this will create a continuing demand for our product and will enable us to survive the approaching crisis.

Barrel sampling with Daniel

How, then, does one attempt to make wine which is significantly superior in quality to the industry average? I have, in the past, been described as "swimming against the tide of the Australian wine industry". I suppose that this is a polite way of saying that I am a heretic! The reader will, by now, have realised that I am heavily influenced by French theory and practice in the production of high quality table wine. Oddly enough, it comes as a surprise to many that irrigation of vines for wine grape production is, by law, forbidden in all French wine producing regions. One may, of course, irrigate if one wishes, but this will entail the loss of the right to "appellation" (labelling of wine in accordance with its vineyard of origin) and a drastic reduction in price. The French are, of course, not averse to making a dollar (or in their case a franc). If they felt that they could improve the profitability of their wine industry by irrigating their vines, would they not do so? Strange then that, to a man, French vignerons are violently opposed to irrigation.

I have discussed the reasons for their opposition with many French producers and their universal view is that irrigation prevents the vine developing an extensive root system and that an extensive root system is an essential component for quality wine production. It seems only logical that vines must, to a degree, be starved of water in order to send their roots deep in search of moisture.

Oddly enough the viticultural gurus of the Australian scene will state, quite bluntly, that it is impossible to grow quality wine grapes without irrigation. How very strange then, that the wines which form the absolute apogee of quality in the Australian industry (Grange Hermitage, Henschke Hill of Grace, Moss Wood and Cullen Cabernets and a Leeuwin Estate Chardonnay) are all produced from non-irrigated vines!

Of course, if vines are to be grown without irrigation, much care must be given to choice of both soil and climate. Sufficient depth of adequately fertile well drained top-soil over retentive sub soils (friable clays) represents the ideal combination. The chosen area should also have annual rainfall approaching 1000 millimetres with reasonable distribution throughout the year but with a relatively dry growing season. Attempts to grow vines in arid climates or in hungry sands over dense clay sub soils without irrigation will, of course, be doomed to failure.

It should be recognised that to maintain vine balance (freedom from excess vegetative vigour) in an irrigated vineyard, cropping levels must necessarily increase. I have a very simple "mental model" in relation to this situation. I believe that each vine is able to produce a certain "quantum" of flavour. One can concentrate this in a small crop of grapes or dilute it over a larger one and the resulting wine will follow suit.

Again, in our winery at Picardy, we tend to use some rather primitive Burgundian techniques, such as whole bunch inclusion in the ferment and hand plunging of the cap of skins on the fermenting must. With his Shiraz, Daniel leaves a considerable amount of yeast in the wine when he puts it into cask. This can make the wine taste and smell quite "feral" in its younger days but will ultimately make it more complex. The "truth in the bottle and the glass" method of analysis, has proven to us time and again, in comparative tastings, that such methods will produce wines of greater flavour, complexity and softness than will other more modern but less arduous techniques.

The next generation of winemakers?

Speaking of Daniel, my relationship with him bears passing comment. Like his father he has a very strong personality and very definite views of his own. He tends to perceive things from his own, sometimes rather unique perspective. Sensing that this could perhaps, create areas of dissension in our relationship, I set out to read everything I could lay my hands on which dealt with the psychology of management. One valuable piece of knowledge which I have gleaned is that similar attitudes and opinions being injected into a management situation will usually result in mundanity coming out of it. Differing attitudes and opinions, on the other hand, will usually lead to radically new and vibrant approaches to problem solving. In simple terms, I have learned to value the differences between us.

Naturally, we have a secret hope that our grandson, Che David, will one day take over Picardy and benefit from the time and effort we have spent in creating it. We do recall, however, that neither Stephen nor Daniel showed great interest in becoming winemakers when we had Moss Wood.

We have recently learned that Stephen has been acclaimed as Red Winemaker of The Year at the 1999 London Wine Challenge. I am afraid I can't really claim any credit for this apart from genetic input. The area in which he works is radically different from the "boutique" situation and I am afraid that I would be totally out of my depth in it.

We have also recently learned that Guillaume Tardy, the son of Jean Tardy, of Vosne Romanee is to spend six months

working with us as part of his training as a winemaker. I expect both Guillaume and Daniel will have much to teach each other. Furthermore, I hope that Guillaume will provide an "entrée" to Burgundy for Daniel.

Lest it be thought that, like my father, I have become "over focused" on one particular aspect of my life, I hasten to assure the reader that I enjoy music (Johann Sebastian Bach, please stand up!), gardening, canoeing and taking my two little canine friends for long, rambling walks. I perceive the hand of the Creator in the majesty of a forest, a dazzling sunset or a night sky packed with stars.

I have practised medicine for some thirty years now and am rather looking forward to retiring from my practice in order to devote myself totally to wine production. I do value and will miss many of the relationships which I have developed in general practice. On the other hand, the level of bureaucratic interference in our activities and the flood of computer generated paper-work have made general practice almost intolerable.

I am of the conviction that I can sleep in one bed, sit on one chair and eat at one table and perceive a fundamental truth in Ghandi's observation that "when mankind equates its needs to its desires, we will have peace on earth".

The older I become the more joy I derive from working with my vines. My tastes are simple and my material needs are few.

Sandra and I feel quite comfortable in Pemberton, much more so than we ever did in Margaret River. There we are literally surrounded by hard-working honest farmers of Southern European extraction. Not only are they industrious but are flexible in their thinking and adventurous in the approach to agricultural activity. This makes a pleasant change from the original Margaret River inhabitants who were, in the main, of British extraction, rather conservative in outlook and resistant to change. (They used to find our viticultural activities quite amusing in the early days!)

Sandra and I find that the attitudes and tastes of the Pemberton community tend to mirror our own, which isn't surprising as we have always preferred French and Italian culture and cuisine to their more staid British counterpart.

And then again, perhaps we ourselves have changed. Perhaps we have matured since our early days in Margaret River, learning to become more tolerant of opposing points of view and to "roll with the blows" to a certain extent. Whatever the reason, we are happy in Pemberton and feel we are there "for the long haul".

CHAPTER 13

VINTAGE 2000

The reader will now detect a subtle change of tense. We are now diarising the present rather than documenting the past.

As I write, we are in the middle of the first vintage of the new millennium. It has, I must say, been rather unusual from several aspects.

According to our eldest son, Stephen, (who works with BRL Hardy in South Australia), the local vignerons have had a very rough time indeed. As Stephen puts it, they have been "hit with just about everything". Early in the season, the Clare district (to the north of Adelaide) was hit by a violent hailstorm which wiped out the majority of their crop. In most other areas, variable climatic conditions, at flowering time, led to very poor fruit set and consequently very low yields. What remains has been sequentially battered by heat waves and drenching rains. All of this has led to widespread outbreaks of fungal disease in the few grapes that remain.

All in all, not a highly memorable vintage with which to begin the new century as far as South Australia is concerned.

On the western side of the continent, we have fared considerably better. Admittedly, we did have some of the same problems with poor fruit set but not nearly to the same extent as out South Australian colleagues; yields being down some twenty-five percent on the 1999 season which was quite a large vintage. In the case of Picardy, this has simply meant that there has been less work in crop thinning the vines (by removing green bunches) in order to achieve the desired cropping levels. In fact, in the Pinot Noir vines, we had to do very little crop thinning at all. (What a blessing!)

We have certainly had more rain than is usual but most of it fell before the grapes began to ripen. The resulting increase in soil moisture has kept the vines in superb condition throughout the ripening period. Probably as a result of this factor together with a low crop load and the absence of extreme heat during the summer, the grapes have ripened with extra-ordinary rapidity. (Vines operate best at temperatures between 20ºC and 25ºC. Above 25ºC, their metabolic function progressively deteriorates and they will "shut down" completely at temperatures above 30ºC).

Another unusual feature of the vintage is the compression of

maturity dates which has occurred. Normally, we would expect the various varieties to ripen sequentially over about a six week interval; commencing with Pinot Noir and finishing with Cabernet Sauvignon. This year, we began picking the Pinot Noir a week to ten days earlier than would be the case in most vintages. The Chardonnay was picked immediately after the Pinot and the Merlot harvested over the following several days. The initial picks of Shiraz were made over the following week. (We have found that grape maturity across the Shiraz block, a distance of some eighty metres, will vary by as much as a full Baume degree. Our practice is therefore, to carefully sample all parts of the Shiraz block and to pick the fruit sequentially, as it ripens to perfection).

We had observed this "compression" phenomenon in the 1997 vintage, wherein it seemed to have been the result of "perfect ripening weather"; a succession of weeks in which day time temperatures hovered between 20°C and 25°C. Interestingly, in both vintages, "compression" of maturity dates was observed right across the continent.

The end result of all this is that the winery was, at one stage, crammed full with crushed grapes and fermenting juices. Every available litre of fermenter space was full and even a variable capacity storage tank had to be pressed in to service as a fermenter.

Daniel: Winemaking is tradition and science

The hectic round of picking, crushing, plunging, cooling and pressing has been quite incessant and Daniel, as the main player, has been understandably quite fatigued by it all.

Sandra and I have tried to help as much as we possibly can and we are all feeling a little "frayed at the edges".

Guillaume Tardy arrived from Burgundy in January and is adapting well to the local climate and conditions; both viticultural and social. He is giving Daniel great support in both the vineyard and the winery. It has taken him a little time to adjust to the fact that we aim to get correct levels of sugar in the grapes in the vineyard and to adjust acid levels (where necessary) in the winery and not the other way around, as is the case in Burgundy.

Daniel does find Guillaume to have a touch of Gallic arrogance about him. As I say to Daniel, a nine century history of producing some of the worlds greatest wines isn't exactly conducive to humility!

Sandra and I find him to be a most charming and affable young man, with a well developed sense of humour. He is living in our new house on the vineyard, cooks and washes for himself and keeps the place spotless. He has quite rapidly developed some firm friendships among the local "blades". He told us recently that he is coming to grips with the local idiom, which, in his view simply requires the insertion of "f---ing" before every noun and verb!

Whatever the other outcomes, the "Pemby Pub" has definitely benefited from Guillaume's sojourn to our country.

On March 10th, we were hit by the remnants of a cyclone which dumped some 25mm of rain on Picardy. The initial heavy downpour was followed by several days of cold, drizzly weather.

Life on a vineyard - always something to do!

This set the sugar levels back (as the vines took up water from the soil) but didn't seem to cause any damage to the grape skins. If the vines absorb sufficient water, the grapes can swell sufficiently to split their skins: leaving them prone to development of fungal disease.

Daniel has developed a technique of spraying sulphur dioxide solution directly on to the grapes, in the vineyard, after rain. This is exactly the same sulphur dioxide which we add to the grapes as they are being crushed and (when used in the vineyard) helps to prevent the development of fungal disease. This allows the grapes to stay clean and healthy while they ripen to desired levels of both sugar and flavour.

In common with winemakers across Australia, we have found that, in order to develop "ripe" flavours in the grapes in the 2000 vintage, we have had to allow them to ripen to sugar levels which are a little higher than those we would normally look for. Normally, we would pick most of the red grapes at about thirteen degrees Baume. In the current vintage, the grapes haven't tasted fully ripe until they have reached thirteen and a half degrees. This possibly relates to rapid sugar accumulation under ideal ripening conditions, with flavour accumulation and softening of grape tannins lagging a little behind.

Our barrels from Tonnellerie Meyrieux arrived well in time for the vintage. If anything, their quality seems superior to those we purchased for the 1999 vintage; possibly because Daniel Meyrieux had more time to source quality oak for the most recent batch. In this year's shipment, we have employed a number of innovative toasting techniques which Daniel Meyrieux (having tasted his wines from our barrels last year) feels could improve the harmony of the marriage between the wine and oak. We have also sourced oak from several different forests. We have oak from the forest of Cher which is adjacent to the forest of Troncais and probably more reliable; the Troncais forest being very small and heavily cut in recent years. We also have a barrel made of oak from northern Spain. This seems to have a character somewhere between French and American oaks and we feel that it could give interesting results when used (in small quantities) with wines from Shiraz.

This year's "mini-disaster" has resulted from our consignment of 1998 vintage wines to the United Kingdom being shipped in an un-refrigerated container; despite our best efforts guarantee that this would not happen. The fact that the loss will possibly be covered by insurance gives us minimal comfort. We had hoped to make a significant impact on the important British market with this, our first significant shipment. Our precious wine, the product of so much planning and effort, will probably wind up on some English rubbish tip and all because of a lack of attention to detail in an area beyond our control. What a crying waste! Be that as it may, the lesson has been learned. Our policy from now on will be to refuse to ship to any overseas customer unless we have a written guarantee that the wine will be transported in refrigerated containers.

From vineyard to glass

In retrospect, the rain which fell after we had harvested the Pinot, Chardonnay and Merlot was a blessing in disguise. It had the effect of slowing the ripening process in the late-maturing varieties (Shiraz, Cabernet Franc and Cabernet Sauvignon) and shifting it in to the cooler autumn weather. The end result has been tremendous accumulation of flavour in those varieties, at quite reasonable sugar levels.

Stephen visited us at the end of March (as part of his regular assessment of BRL Hardy/Houghton's fruit sources) and was excited by the quality of the Cabernet Franc grapes at Picardy. In his opinion, they were superior to all examples of that variety which he had seen elsewhere in Australia.

We, ourselves, are quietly excited about the quality of the wines we have produced in the 2000 vintage. Quantities are certainly down but quality is way, way up.

The last of the Shiraz and the Cabernet Franc were picked on April 1st (not, one hopes, prophetically) and the Cabernet Sauvignon on April 10th. In the fermenter, the Cabernet Franc had a bright red colour. This rather surprised us as its colour is usually more of a "muddy" brown.

"Gill" Tardy is finding the length of our vintage a bit of a problem. In Burgundy, their harvest only lasts several days. When we tell him that some large Australian wineries can be in vintage for up to five months. He finds that difficult to comprehend!

Sandra, Jodie, Daniel and I tasting the wares in the Picardy barrel room

Whereas vintage 1998 brought a new life to our family (in the person of our first grandchild), that of 2000 took one away. Our little canine companion of eighteen years, R.D. Bollinger left us on 28th April. Only a dog, you may say, but for us he was very much part of our family. He had been with us since the latter days of Moss Wood, throughout our involvement in Smithbrook and in the formative years of Picardy. He had helped to raise all of our children, giving them something to care for in their developing years. For a creature of quite diminutive stature (being a miniature Dachsund) he was unbelievably courageous and had his own unique little personality. Sandra and I are missing him terribly.

By a series of unrelated circumstances, those of our family are no longer in Western Australia have congregated in South Australia. Stephen is based at Reynella, south of Adelaide. Our third son, Nicholas, having completed a commerce course at the University of Western Australia decided that the business world was not for him and joined the army. He has recently been posted to Adelaide. Nicholas feels that three Pannell males is about as much as the Australian wine industry could tolerate and has no ambition to become involved therein.

Our youngest child and only daughter, Emma, who spent most of her infancy in a cot or play-pen at Moss Wood, swears that she never wishes to see another vineyard or winery for as long as she lives. Her partner, Damian, is now in the third year of a wine-making course at Adelaide University. Poor darling Emma has not, it seems, been able to escape the Pannell family's vinous affliction! Fate does indeed work in strange ways.

ET CETERA

It will be obvious to the reader that some little time has elapsed since I first put pen to paper in order to record these, my memoirs.

We have had a further ten vintages since the turn of the millennium and two close friends (Tom Cullity and Ellis Butcher) have left us.

I will lay the blame for the delay in publication at the feet of "regulation" publishers who, when shown the manuscript in 2000, demonstrated a marked lack of enthusiasm for (and a marked lack of interest in) it.

More recently, the development of desk-top publishing together with the encouragement and assistance of Peter Rigby (not to mention the inevitable approach of my own demise) have stirred me, once more, to attempt to present my narrative to the interested public. For the sake of completeness, I should present a brief synopsis of those vintages which have occurred since the year 2000.

2001 - A rather warm year which produced excellent Chardonnay and late reds (serious Bordeaux blends). While very drinkable, the Pinot Noir lacked a little elegance and complexity.

2002 - A marvellous vintage across all varieties. A small crop and perfect ripening conditions produced great wines in all categories. The 2002 Pinot Noir and Bordeaux blend (Merlot/Cabernet Sauvignon/Cabernet Franc) are both drinking superbly now.

2003 - Another very warm year and average crops produced good (but not great) wines from Pinot and Chardonnay. The late reds were, however, of very high quality. Despite our own reservations, the 2003 Tete de Curvee Pinot Noir won both sections (with and without food) of the Australian Pinot Challenge in 2006: and achievement not previously accomplished by the same wine.

2004 - What could have been a high quality year was marred by the relevant government body conducting a pre-emptive burn in the forest close to our vineyard in late December 2003. While in barrel, the wines showed no evidence of smoke taint, this did become apparent (albeit at a very low level) after the wines were bottled. Fortunately the Chardonnay wines from that year were uneffected and are still drinking superbly to this day.

2005 - A marvellous vintage in which a medium crop ripened under near perfect climatic conditions. More frequent plunging (pigeage) of the Pinot ferments (on the advice of a Burgundian friend) produced wines of great intensity and complexity.

2006 - Certainly the most difficult year (climatically speaking) which I can recall in my entire career in winemaking. The Pinots and Chardonnays (being early ripeners) achieved adequate maturity but persistent cool and rainy conditions mitigated against adequate ripening in the late reds. Picardy did not release a Shiraz or Bordeaux blend from 2006.

2007 - A very good year, climatically speaking. Higher than average bunch weights meant that yields were rather higher than intended. As a result, the Pinot lacks a little concentration. The Chardonnay and late reds are of excellent quality.

2008 - By way of contrast with 2006, this was the best vintage I have ever been involved in. Across all varieties, yields, maturity levels, concentration of flavour and acidity levels were exactly as desired. The resulting wines have "tightened up" a little in bottle (as so often happens with great vintages). We are confident that they will develop into some of the greatest wines that have been produced in this country.

2009 - A small crop which ripened under excellent conditions. In barrel, all the wines showed great concentration and complexity.

2010 - Average yields of high quality fruit. A hot spell in late February produced Pinot and Chardonnay of great concentration. A cool March meant that the late reds took at least another month to ripen and have produced wines with supple tannins and ripe flavours.

On a personal note, I retired from medical practice in 2001 and am enjoying the luxury of devoting my time entirely to Picardy. I remain in good health and am very much involved in vineyard and winery activities on a day-to-day basis.

In 2003, I was awarded membership of the Order of Australia for my services to the wine industry.

What of the future for Picardy? In both the vineyard and the winery, we adhere to a policy of incremental improvement in all aspects of the production process.

New Pinot Noir planting material has been sourced from a friend in Burgundy and will be released from quarantine in 2011.

Whilst we are a little skeptical regarding the claims made (by some producers) for the organic and biodynamic approaches to wine production, we adhere to a policy of sustainable agriculture and continuous improvement of soil health in the vineyard. To this end, we avoid the use of insecticides and heavy application of chemical fertilizers.

On the other hand, we are strong believers in the benefits of natural manures and composts, together with agents which stimulate soil microbial activity. We like to think of our approach as "scientifically based biodynamics'".

Be that as it may, it is impossible to make good wine from moudy grapes and the use of some systemic fungicides in the cool, damp climate of Pemberton is (in this context), unavoidable.

Sandra and I try to visit Burgundy as often as possible; usually once each year. During the course of our stays, we invariably manage to pick up new and interesting ideas. Frankly, I find that these visits "recharge my batteries" and I return to Picardy with renewed energy and enthusiasm.

The one factor which our Burgundian friends universally regard as a "sine qua non" for quality wine production is *vine age*. Modern technology notwithstanding, it still takes 365 days for a vine to become one year older. The consensus seems to be that vines will begin to produce wines of real quality from about 20 years of age.

Our oldest vines now are 17 years behind them. God willing, I will live to see what really old vines are capable of producing at Picardy.

First pick and riders of the Pemberton Classic 2010 zip by Picardy

Seasonal moods at Picardy.

Daniel and I destemming the Pinot

Exhibiting at Wine Australia, Sydney

John and Attilio at the winery

With my long-time friend Max Lake after judging Pinots in Sydney

My Order of Australia award

Premiers's Hospitality & Tourism Award

Dan and Jodie with the prize magnum at a Picardy tasting

WA WINE PRESS CLUB PRESIDENT'S NIGHT

INCLUDING

PRESENTATION OF JACK MANN MEMORIAL MEDAL

BY

HIS EXCELLENCY the GOVERNOR, Lt-Gen. JOHN SANDERSON, AC

IN THE

GOVERNMENT HOUSE BALLROOM, ST GEORGE'S TERRACE, PERTH

ON

TUESDAY, 28th MAY 2002

Guests to arrive between 7 and 7.30pm

Dress: Black Tie or Lounge Suit

WA Wine Press Club Award and Presentation of the Jack Mann Memorial Medal

APPENDIX

One of our proof-readers has made the perfectly reasonable observation that, with all the discussion of grape growing and wine making in the forgoing pages, he would like to gain a greater understanding of the actual wine making process.

Techniques involved in the production of fine wine can, with all their subtleties and permutations, appear extraordinarily complex. The basic, underlying process (the conversion of grape sugar into alcohol with the retention of flavouring substances and in the case of red wines, colour) is however, eminently comprehensible.

In the interests of informing rather than confusing the reader, I intend to adhere to the "simplistic" approach in the ensuing discussion.

It goes without saying that, in order to produce quality wine, one must begin with quality grapes. Contrary to opinion in certain sections of the Australian wine industry, silk purses are rarely produced from sow's ears!

Ripe grapes contain six major components; water, sugar, acid, flavouring substances, tannins and colouring substances or pigments.

The water content will vary both with the maturity of the grape, the seasonal conditions (excessive heat or rain at vintage) and the manner in which it has been grown (irrigated or non-irrigated).

The sugars (glucose and fructose) accumulate in the berries as they ripen and at maturity, will constitute a little over twenty-two percent of the grape's juice. Such levels will usually result in alcohol levels at about thirteen percent in the finished wine.

The acids (essentially tartaric and malic) become less concentrated in the juice as the grape ripens and will, at maturity, be present at concentrations of between six and eight grams of acid per litre of juice.

The flavouring substances accumulate as the grape ripens and are usually unique to a particular variety but in some cases, are shared between varieties eg. Muscat and Traminer.

The tannins are mainly present in the skins and seeds. They belong to a group of chemicals known as phenolics and have an astringent or slightly bitter taste.

The pigments are almost invariably present in the skin of the berry and belong to a group of chemicals known as anthocyanins (which are closely related to the tannins or phenolics).

At maturity, the grapes must, above all, have adequate flavour. Jack Mann's son, Dorham makes the rather sweeping statement that one can add anything to grapes but flavour. In essence, he is correct but ripe grapes should also be "in balance"; which means that the relationship between sugar, acid and flavour should conform to certain parameters if the resulting wine is to be fresh, balanced and interesting to drink.

Certainly one can add water and acid to over-ripe grapes or (in Europe but not in Australia) sugar to under-ripe grapes. Acid can also be added to grapes deficient therein but otherwise in balance. Nevertheless, wines produced from grapes which are in "perfect balance" with respect to sugar, acid and flavour, will always be superior to those which have required "adjustment" with respect to their levels of sugar or acid. Over-ripe grapes tend to produce wines with "stewed" or "jammy" characters and excess acid addition leads to a "hardness" in the resulting wines.

Our policy at Picardy is to pick on flavour; that is, when the grapes taste ripe but not over-ripe. Fortunately for us, in the cool climate of Pemberton, this usually means that the grapes are otherwise in balance with sugar levels at about thirteen degrees Baume (23.4 percent sugar) and with acid levels in the range of seven to eight grams per litre of juice; an ideal situation from the point of view of producing balance in the finished wine.

At Picardy, we also pay considerable attention to juice pH. pH levels are generally (though not invariably) inversely related to levels of acidity. Low pH levels in the grapes result in finesse and stability (good keeping qualities) in the resulting wines. It is our good fortune to be able to obtain extremely low pH levels in our grapes at quite moderate levels of juice acidity.

Once ripe, the grapes are picked and transported, as quickly as possible, to the winery where they are passed through a machine which removes the berries from their stalks (by beating them through perforations in a rotating cylinder) and then crushed between rotating rollers. Older machines used to crush the berries prior to removing the stalks but it is now considered that this led to extraction of bitter tannins from the stalks and the former process is preferred nowadays. A little sulphur dioxide (preservative / anti-oxidant) is generally added to the grapes during the crushing process.

At this stage, the "streams" for the production of white wines and red wines diverge.

For white wines, the crushed grapes are pumped directly into a press. The oldest and simplest presses take the form of a wooden cage or "basket" which is filled with crushed grapes. Heavy wooden blocks which fit precisely into the top of the basket ("head boards") are then forced down on to the crushed grapes and their juice is forced out between the wooden slats of the basket. More modern presses generally employ an inflatable bag contained within a perforated (or "slotted) stainless steel cylinder. The crushed grapes are fed into the space between the bag and the cylinder and the bag is inflated, resulting in the juice being forced through the perforations in the cylinder.

Whatever the pressing technique employed, the grape juice is collected in a large tray and then pumped into a holding tank. It is chilled to a low temperature and its acid level adjusted, if necessary. Over the next several days, solid material in the juice (grape pulp, skins, seeds etc.) will settle to the bottom of the tank. The clear juice is then pumped off and transferred to another tank where it is inoculated with a suitable yeast. Within several days, fermentation commences. This process converts the grape sugars into alcohol in accordance with the following chemical reaction: C6H12O6 (sugar) ➤ 2C2H5OH (alcohol) + 2CO2 (carbon dioxide) + Heat

As can be seen from the above, the fermentation process is "exothermic" (produces heat") and the fermenting wine will rapidly heat up if it is not continuously cooled, (usually to a temperature of about 20 degrees Celsius). This will damage the wine by driving off many of the low boiling-point, volatile flavouring substances and will produce a wine which is lacking in flavour and freshness if the ferment is allowed to overheat.

The wine will continue to ferment until all the grape sugar is converted into alcohol; usually over a period of several weeks. Some styles of white wine (eg. Semillon, Riesling) are produced entirely in stainless-steel tanks. Others, notably Chardonnay, are matured (and sometimes fermented) in oak barrels, in order to make the wine more complex and to impart a desirable level of oak flavour thereto. Not infrequently, Chardonnay wines are allowed to rest on their yeast sediment (or lees) to further increase their complexity. Regular stirring of the lees ("batonnage") will further enhance this process.

When the white wine is judged to be ready for bottling, it is run off its yeast sediment and transferred to a clean tank. It is "fined" (with Bentonite clay, isinglass or similar) to remove unstable proteins which might otherwise form a cloud in the bottle. The wine is then "cold-stabilised" (held at a very low temperature for an extended period in order to remove unstable cream of tartar crystals), filtered and bottled.

In the case of "tank-matured" white wines such as Semillon and Riesling, bottling may be carried out within weeks of the completion of fermentation. With barrel matured wines such as Chardonnay, they may remain in cask for up to a year prior to bottling.

As has been previously stated, the colour pigments in red grapes are found almost entirely in the skin of the berries. It therefore stands to reason that in order to produce red wines, those pigments must be extracted from the skins.

After crushing, the juice, skins, pulp and seeds are pumped directly to a fermentation vat. Acid levels are, if necessary, adjusted and a yeast culture is added. The fermentation process commences shortly thereafter (the crushed red grapes or "must" being at a rather higher temperature than the clarified white juice).

In contrast to white wine fermentation, where the carbon dioxide produced escapes easily from the clarified juice, the escaping gas in red ferments tends to be trapped by the grape skins which are thus lifted to the top of the ferment where they form a dense "cap". Evolved heat also tends to accumulate in the cap and (in the absence of corrective action) will reach temperatures which will damage the yeasts and foster the growth of undesirable organisms, leading to the production of hydrogen sulphide (rotten-egg gas) and acetic acid (vinegar acid). It is therefore imperative that the floating skins be broken up and submerged in the fermenting wine at least several times each day. This can be achieved by pumping the juice from the bottom of the fermenter and spraying it over the cap ("pumping over") or by the more laborious technique of "plunging" the cap with a flat plate in the end of a long pole. At Picardy we prefer the latter method as we believe that pumping over leads to the extraction of coarse/bitter tannins from the skins and seeds.

"Roto-fermenters" (which resemble concrete mixers) are gaining popularity in large wineries as the cap can be broken up and mixed with the juice by simply rotating the fermenter. If used injudiciously, these can lead to excessive extraction of tannin.

Again, the fermentation process is exothermic and the fermenters must be continuously cooled throughout fermentation, though temperatures for red ferments tend to be considerably higher than those for whites: being in the order of 28-30 degrees Celsius.

The combined action of heat, alcohol and agitation of the cap (by plunging or pumping over) leads to the extraction of red pigments (and tannins) from the grape skins and provides colour in the resulting wine.

At the completion of fermentation, the juice, skins and pips are transferred to a press where the juice is separated from the solid materials.

The young wine is then stored in a holding tank for several weeks. During this time, the yeast cells and other solid matter will settle out of the wine and fall to the bottom of the tank. The clear wine is then pumped into barrels where it will remain for the next one to two years, (depending upon the style of wine which the maker is attempting to produce). During its time in barrel, some of the wine will evaporate and the barrels will need to be "topped up" at regular intervals.

From time to time, the wine may begin to taste stale or

excessively yeasty and at such times will require "racking". This involves removing clear wine from the barrel, washing out any sediment which as formed in the bottom of the barrel and then returning the wine thereto.

During their time in barrel, red wines (and some whites) will usually undergo a secondary fermentation in which unstable malic acid is broken down, by bacteria, to the more stable lactic acid; a process involving the production of more carbon dioxide. If this were to happen in bottle, the red wine would be spoiled (becoming both effervescent and cloudy). Little wonder, then, that the "malo-lactic fermentation" is actively encouraged during the wine's time in barrel, by all makers of red wine.

Other subtle changes occur when red wine is stored in oak barrels. In addition to absorbing desirable oak flavours, the wine undergoes a very slow, controlled oxidation which leads to the production of complex flavours and softening of its tannins. (The chemical structure of tannins resembles short chains which have the ability to "polymerise" or join together to form longer chains. "Short chain" phenolics have a very bitter taste but the longer the chain becomes, the less bitter is its flavour. This polymerisation of phenolics, which occurs naturally over time is, in fact, the chemical basis of ageing in red wine).

At the end of their period of maturation in oak, red wines are fined (usually with egg-white or gelatine) filtered and bottled.

Unlike white wines, which are often sold quite soon after bottling, reds usually require a period of maturation in bottle in order to recover some of the flavour which is usually lost during the bottling process ("bottle-shock") and to allow the tannins to soften even further.

Picardy, Pemberton

AKNOWLEDGEMENTS

- To the late Jack Mann who provided the spark that lit the fire.
- To the late Bill Jamieson (State Government Viticulturalist), for his invaluable assistance in the early days at Moss Wood.
- To the late Tom Cullity for his cameraderie in those early years.
- To Gill Thomas for his companionship and support at Moss Wood.
- To Becky Wasserman and Russell Hohn of Bouilland in Burgundy, who opened many doors for me and led to my involvement in Domaine de la Pousse D'or.
- To the late Henri Jayer of Vosne Romanee in Burgundy. In a one-on-one interview and tasting in 1982, he developed in me a passion for Burgundian-style winemaking.
- To Gerard Potel (and numerous other Burgundian winemakers) who reinforced that passion.
- To Carel Verhuis of Domaine D'Ardluy, Corgoloin, Burgundy, who taught me about clones.

Sandra's favourite garden water feature at Picardy